YOUR ULTIMATE GUIDE TO ABUNDANT ENERGY

Brad J. King

Health Venture Publication

Health Venture Publications
5948 3rd Line RR#1
Hillsburg, ON
N0B 1Z0

Cover Design: Sam Truax
Book Design: BbM Graphics
Copy Editing by: Bruce W. Cole,
Cole Communications

Ultimate products were developed by Brad J. King, M.S., MFS author of *Fat Wars* and the *Fat Wars Action Planner.*

Disclaimer.
It is very important for you consult a doctor before making changes in your diet and lifestyle, or before taking vitamin and/or food supplements.

While all care is taken with the accuracy of the facts and procedures in this book, the author accepts neither liability nor responsibility to any person with respect to loss, injury or damage caused, or alleged to be caused directly or indirectly, by the information contained in this book.

The purpose of this book is to educate and inform. For medical advice you should seek the individual, personal advice and services of a medical professional.

contents

contents

In order to conserve pages, references may be found on:
www.awakenyourbody.com

introduction

JUST CHILLIN' WITH MY KIT CAT

ASK ANY TEENAGER how old they are and they will answer you faster than you can finish the question. Why? Because we simply don't connect any negativity to age when we are teenagers – other than the things that matter most, like being old enough to get our driver's permit and the most important one of all, finally reaching the legal drinking age!

Ask someone in their thirties how old they are, and you'll have to wait a little longer for the reply. Now, try the same experiment with a woman in her forties or fifties (women are a little more sensitive to this issue) and I can almost assure you that her response would be something like; "It's none of your damn business." But why? The answer is simple. The average North American today does not like what they see and/or experience after that "certain age". What "certain age" you ask? Well, that depends on how your individual biochemistry has held up against the onslaughts you have dealt it over the years. But the majority of North Americans past the age of forty today are not exactly happy with the way their bodies look, feel and perform.

Now you're probably thinking that past middle age we can't exactly expect to look, feel and perform like a teenager and that it's normal to experience the dreaded decline that seems so inevitable with advancing age. The problem is that it has become normal in this day and age to lose muscle mass, strength, bone mass and our once-vital energy potential. It's also normal to notice body fat increases, wrinkling of the skin, various mood changes and a declining immunity (usually experienced as more frequent colds, aches and pains). But just because something is perceived as normal does not mean that it is the way it should be! Normal certainly does not reflect healthy – at least in North America! It just means that the majority of us seem to accept this decline in vitality as the way it is *supposed* to be as we get older. Never strive for normalcy.

You were once young and vital and with that came seemingly unlimited energy.

As I write this, my nine-week-old kitten, Kit Cat (yeah, I know I'll receive flack for this one), is continuously tapping at my keyboards, using my leg as a climbing post and doing everything in her power to persuade me to play. Why? Because her cellular energy cycle is currently flawless (but by the scratches on my legs, certainly not clawless!) This brings me to the overall concept of this book: aging, to a very large degree, is caused by a reduction in our capacity to generate energy at the cellular level. This loss of energy output, which you will learn about in the following chapters, creates a progressive decline in the way our bodies function and an unprecedented susceptibility to age-related diseases.

But new science shows that it doesn't have to be this way. According to myriad research, what separates a young cell from an old one is the cell's ability to churn out energy (also referred to as metabolism) and protect itself from the accumulative damage from daily living (i.e. the air we breathe, the food we eat, and our lifestyle choices). It is the cell's ability to continually repair itself and maintain healthy cellular renewal through metabolism that allows us to experience youth as we know it.

As I work to complete this introduction (which is taking much too long due to a certain creature attacking my leg again), I can't help but be amazed by what is behind Kit Cat's limitless energy potential. One cannot help but be humbled by the driving force that propels us through life and the perfection of cellular orchestration that takes place in youth. And all these trillions of biochemical reactions that allow us to do everything we take for granted – including my facial expressions every time Kit Cat climbs up my leg, take place without so much of a moment's thought by us.

You were once no different from Kit Cat in your ability to create youthful vitality. You also once experienced limitless energy when you were young, or at least when your cells were. And that energy was always there for the taking. But then one day, you woke up and the energy was not what it once was. The saddest part of all is that, like almost everyone else, you eventually accepted this less than optimal energy production as your new way of life. It's as if you went into work one day to find out that you had been demoted from a job you loved, to one that was less fulfilling and didn't allow you to experience life the way you had become accustomed to. What you should have done was march right up to the head honcho and fight for your old position back but instead, you simply accepted your new status as if there were nothing you could do about it. But the saddest part of all was you never even made an effort to regain the life you had loved.

And so it is with the majority of people in this day and age. They sit back and do nothing about their present physical situation, they make the wrong choices, or they simply don't know what to do.

In *Awaken Your Metabolism*, I am going to share with you some incredible research that paints a very different picture of what aging and energy can represent to you, on a biological level, if you are willing to take the necessary steps to awaken your youthful vitality. I'm not talking about stopping the aging cycle, but I am going to show you how you can create a metabolism that works in your favour as you age, a metabolism that allows your body to become younger where it counts most – in your cells. If you are open to a new understanding of aging, one that isn't affected by chronological years or candles on your birthday cake, you are in for a real awakening.

Yours in Abundant Health

part one The Magic Inside

CHAPTER 1

HOW DO YOU CURRENTLY LOOK, FEEL AND PERFORM?

Caution! You are about to embark on a journey that will lead you away from a warm and fuzzy place, a place called "your comfort zone." This is where a great deal of the population spends the majority of their lives, doing exactly what they have been doing since they can last remember and unfortunately experiencing a life of little biological potential because of it.

The biological potential I speak of lies in your body's intrinsic ability to remain in optimal health for the great majority of your life here on earth. The point is, every single metabolic reaction responsible for how you feel every second of every day is controlled through metabolism at the cellular level. That's right, every cell in your body has its own metabolism, and if treated right – that is provided with the right fuels and lifestyle choices – that metabolism will allow your trillions of cells to work in harmony with one another. The outcome, simply enough, is a life lived with abundant energy and vitality and free of disease. Now, why is this concept so hard to believe in this day and age? Don't you think you deserve to live your life this way? I hope you said yes.

The question is, why are you not feeling this way now? Too many of us learn to live with broken bodies that arise from inefficient metabolisms. It's as if we accept this metabolic decline as we get older as the way it is supposed to be. Don't think for one second that your body is designed for anything less than the best performance it can give you. But understand that, like everything else, if you continuously abuse the system, it will have no other choice but to eventually break down and self-destruct.

This is exactly what the majority of North Americans experience as they get a little older, in the biological sense, that is: a body that becomes less and less efficient at maintaining the daily wear and tear of living. The outcome is less energy, more girth, memory deficits, aches and pains, leathery skin, loss of strength, greater susceptibility to illness and eventually early death.

I'm not trying to paint a picture of desperation for the human race here, I am simply stating that the human body cannot expect to be maintained in an optimal fashion if it is not given the necessary tools to accomplish this state.

This book is designed to give you the tools that you will need to rebuild your body's metabolic potential. If you apply the information within these pages – you will not only improve your present health status and regain your true life potential, you may actually rewind the clock a few years in the process!

So many of us have forgotten what it means to be alive and to live at our fullest potential. Instead, we become mere spectators that sit out the game of life, yet still have the audacity to stand in front of that full length mirror scratching our heads and asking, "What the heck happened?" It's as if we are watching a film with us in the lead roll and not knowing what direction the storyline will take. Worse yet, many people don't even know what their story is about. I have news for you. You are not only the star in your movie, you are the director and producer and therefore you control the movie's destiny.

Remember what it was like when you were young? Remember getting up in the morning as a kid, flinging opening the blinds, scarfing down your breakfast (even sugar-coated processed cereals kept you going back then) and then hopping on your bike for the entire day? Weren't you happy to be alive? How could you not be? How many of us talk about the things we were able to do in our youth? Things we unfortunately took for granted back then and all because they always seemed to be at our beck and call.

Now as adults, most of us get up out of bed grumbling, half-awake, grab a couple of donuts and a double espresso, and spend the entire day gulping coffee in offices with circulated stale air. Then, after a hard day at work, we come home and eat most of the day's calories in refined carbs while watching TV on the couch. Things have certainly changed, wouldn't you say?

The information you are about to read in no way contains the map to the Fountain of Youth, nor does it contain ingredients to a magic potion that will help you revert back to that child you once were. The truth is, there is no instant gratification to a life worth living. Having said this, if you follow the advise within these pages, you can become younger – biologically speaking. I'm obviously referring to "youth" in the sense of how effective your metabolism used to be at using fat for energy, how supple your skin was how quick your mind was, and let's not forget how great you looked, felt and performed.

"Remember what it was like when you were young? Remember getting up in the morning as a kid, flinging opening the blinds, scarfing down your breakfast and then hopping on your bike for the entire day? Weren't you happy to be alive!"

Being young is not objective, it is biological, and it can be measured scientifically with various biomarkers that are proven to access how old your body is from the inside out. It just so happens that one of the most, if not the most, important biological marker of how we look, feel and perform lies in our ability to maintain *lean body mass* or muscle. In fact, a study of 84 men and women aged 90 to 106, presented in the *Journal of the American Geriatrics Society* in 1997, showed that loss of muscle is the number one determining longevity factor – that is, in shortening life span.[1]

In my book *Fat Wars*, I explain that, to a large extent, muscle controls the overall metabolic rate of the body.[2] And in case you forgot, your metabolism is responsible for how much energy you produce, not to mention how much of that energy comes from your very own fat stores!

The good news is that the information provided here will give you the keys to correct and maintain an optimal metabolism. I will show you that it is not only possible to grow lean muscle tissue at any age, but that your body also holds the ability to reactivate your existing (often stagnant) muscle, ultimately giving you a more youthful metabolism in the process. This equates to a more energetic, leaner and healthier you!

I'm not saying that looking good is all that counts, although according to the cosmetic and garment industry, it happens to be a major motivating factor in most people's lives! What you need to realize is that a declining metabolism is one of the dominating factors behind a less than optimal health status, not to mention the new "spare tire" many of us carry around our waist. The fact remains that every single extra ounce of energy you lose and every pound of body fat that creeps up on your unsuspecting body does a lot more than make you feel a little less energetic or attractive. As you will see in the next chapter, it slowly erodes your life potential.[3]

CHAPTER 2

WHAT IS METABOLISM, ANYWAY?

Mention the word "metabolism" at a dinner party and you're likely to receive blank stares from some, while the rest of your friends straighten up and pull in their stomachs to look thinner. Many people either have no clue what metabolism means, and many more simply believe it has something to do with weight loss. While your metabolism is an important factor in your body shape, the strength of your metabolism affects the health of every single cell in your body. From "*metabol*" meaning change, metabolism refers to the chemical reactions that help cells transform energy so they can use it to remain healthy and reproduce.

When we're discussing metabolism, we are referring to both our "total metabolism" (all of the biochemical processes in our bodies) as well as "cellular metabolism" which includes all the chemical process that take place in our cells. When you consider that our bodies contain approximately 100 trillion cells – all requiring equal attention, you get an inkling of the importance of having a healthy metabolism. Poor food choices, (too much of anything or not enough, or that last fad diet), sudden over-exertion, repetitive physical activities or too little rest – all these things generate little assaults on each and every cell that makes up your body. These assaults may not seem like much at first, and they may even go unnoticed for years, but eventually, the damage accumulates to affect muscles, organ systems, your bones, and your skin, until finally, you are feeling more like your "old self" than your young self.

Life's not-so-little Balancing Act

Metabolism involves two processes: catabolism, in which larger substances are broken down into smaller substances, and anabolism, which refers to building larger structures from smaller ones. So, for example, a mouthful of dinner contains large particles that are broken down through catabolism to the smaller nutrients that your body is able to use for fuel. Through anabolism, those smaller nutrients are used to build and repair the cells of your bones, muscles and skin etc. When you think *anabolic*, think "repair." The more anabolic our metabolism is, the more youthful we look, feel and perform. For instance, when we are young and highly anabolic, our body repairs and replaces nearly 300 billion cells each and every day. As we age and the everyday wear and tear of living starts to get the better of our metabolism, we become more *catabolic*. Catabolism is also responsible for dismantling old and damaged cells to prepare them for elimination from the body. Catabolic means "break-down." In order to experience a healthy metabolism, we need to strike a healthy balance between these two processes (anabolism and catabolism), with the strengthening forces of anabolism being slightly more active.

"As we age, we often don't do the things that promote a healthy metabolic balance. We don't exercise enough, we don't eat the right balance of foods, and we forsake sleep for late night movies or to catch up on work from the office."

When we were young, anabolism ruled. Think again about those summer days on your bicycle when you took in plenty of sunshine, fresh air, and most likely, food prepared in our very own kitchen by your mother or grandmother. But gradually, catabolism slowly starts to pull ahead until one day, "WHAM," it has taken over. That's because, as we age, we often don't do the things that promote a healthy metabolic balance. We don't exercise enough, we don't eat the right balance of foods, and we forsake sleep for late night movies or to catch up on work from the office. Yet proper food, exercise and sleep are essential to keep our metabolic processes in check. With a lifestyle of fast food and slow physical activity, we get a whole lot more body break-down without the right nutrients to support the rebuild and repair functions. In other words, aging is very much a process of becoming ever more catabolic.

Denying our bodies these metabolic necessities opens the door to multiple consequences, including but not limited to low energy status, muscle wasting, bone loss, inflammatory conditions, hormone dysfunction, nervous system disorders, skin conditions and cancer. And you thought your metabolism only controls how much you weigh!

Taking it to the Cells

Inside the majority of your trillions of cells lie numerous little energy plants called mitochondria, tiny sausage-shaped organelles that constantly move and change shape. A combination of enzymes and oxygen within the mitochondria help to break down foods to release energy. While much of this energy escapes the mitochondria as heat, these tiny energy factories are responsible for producing at least 90 percent of the energy-carrying substance adenosine triphosphate (ATP). ATP is extremely important, because it provides a form of energy that can be used by every cell in the body. In fact, you produce almost 200 pounds of ATP every single day.[4] And while glucose is the most important fuel in the body, it can't directly be used to perform cellular work. Instead, glucose is catabolized and stored in ATP to be used as an energy source when required. Among other things, ATP helps amino acids to cross cell membranes, helps muscles to contract and lengthen and provides the resources needed to promote chemical reactions that help to absorb energy.[5] So you can see the importance of keeping your mitochondria in top working order. And we do that by fueling our metabolism properly.

Measuring your Metabolic Power

Besides the obvious excess weight that we might be carrying around, many people aren't aware that their metabolism might need a boost or a metabolic tune-up. But while there might not be any obvious symptoms, heart disease, diabetes, arthritis and other chronic conditions don't just happen overnight. So, how can you tell how your metabolism is doing?

To answer that question, scientists have devised the Basal Metabolic Rate (BMR). This figure measures the amount of energy (calories) you burn simply by being alive, including heart-beat, breath and other physiological functions. Because testing for BMR must take place in a research facility following a 12-hour period in which you do not move or eat, it is rather difficult and inconvenient to measure. As a result, scientists now use the Resting Metabolic Rate (RMR) to measure effectiveness of the metabolism. The RMR includes the BMR, plus the extra energy required to be awake and alert. Because the RMR accounts for approximately 60-75% of the calories you burn daily, it makes sense that the higher the RMR, the more effective your metabolism is at converting food into energy, all the while allowing you to burn gobs of body fat and maintain a lean, healthy physique. The RMR is closely linked to your percentage of lean muscle mass, and the more lean muscle you have, the higher the RMR.

Body Mass Index

Because measuring your RMR is not something you can do at home, you can also use your body mass to gauge the effectiveness of your metabolism. Your body mass index (BMI) is calculated with a formula that considers both your height and your weight. Your BMI is determined by your weight in kg divided by your height in metres squared. [For instance, if your height is 1.68 (5'6") metres, the divisor of the calculation will be (1.68 x 1.68) = 3.36. If your weight is 68.4 kilograms (150lbs.), then your BMI is 20 (68.4 divided by 3.36).]. The ideal BMI falls between 19 and 25. A BMI higher than 25 puts you at greater risk for health problems associated with obesity.

While the BMI is a useful yardstick for measuring your metabolic potential, it is not a perfect system because it doesn't distinguish between men and women, or athletes whose healthy muscle mass might put them in a higher weight category. Still, it's a good place to start.

Waist Not

Another important predictor of disease also seems to be the size of your spare tire. Strongly associated with abdominal fat, waist circumference reliably predicts weight-related disease, including cardiovascular disease and diabetes. A measurement of more than 35 inches for women and more than 40 inches for men signifies increased risk of disease. For accurate results, measure your waist at the midpoint between the hipbones and the bottom of the rib cage while standing. A measurement of 35 inches or more for women and 40 inches or more for men indicates a risk for cardiovascular disease and diabetes.

Why Metabolism Matters

It's important to remember that chronic conditions like diabetes and serious diseases like cancer don't just occur overnight. The road to illness can be long and winding, and our bodies constantly send us clues about the state of our internal health. Unfortunately, more often than not, we aren't paying attention. But we can do much to preserve our well-being if we keep our metabolic fires burning. Your trillions of cells will continuously reward you with abundant energy, health and vitality. Now doesn't that sound an awful lot like what you experienced when you were a kid?

"Health is like money, we never have a true idea of its value until we lose it."

– Josh Billings

CHAPTER 3

THE HORMONE CONNECTION

Although hormones are most often associated with reproductive issues, these power chemicals play a role in everything from thyroid health to hair loss. Of course, many of the metabolic messages in your body are also controlled by your hormonal system.[6] Hormones are tiny little protein or cholesterol-derived chemicals that scurry through the blood stream translating biological messages to each of your cells. Once a cell picks up these messages, various catalysts called enzymes are awakened to carry out the actual orders. Think of hormones as the "kingpins" of the organization and enzymes as the ones that get their hands dirty.

When it comes to the message of fat storage, for example, the hormone insulin plays the leading role. Excess insulin caused from dietary imbalances stimulates the production of an enzyme that sits atop your fat cells called lipoprotein lipase or LPL. (*See Figure 1 next page*) The more LPL you stimulate, the fatter you become.[7] Fortunately, cells also contain a fat-releasing hormone called glucagon. Glucagon, in turn, stimulates the production of another enzyme that sits atop your fat cells called hormone sensitive lipase or HSL.[8] (*See Figure 2 next page*) As LPL creates continual fat storage chemistry, HSL does quite the opposite, allowing for continual fat release and energy production.[9,10,11] Trust me; HSL is one enzyme you want on your side. The good news is that through the proper combination and timing of foods, you can increase levels of the fat-releasing hormone glucagon,[12] and thereby its subordinate HSL.

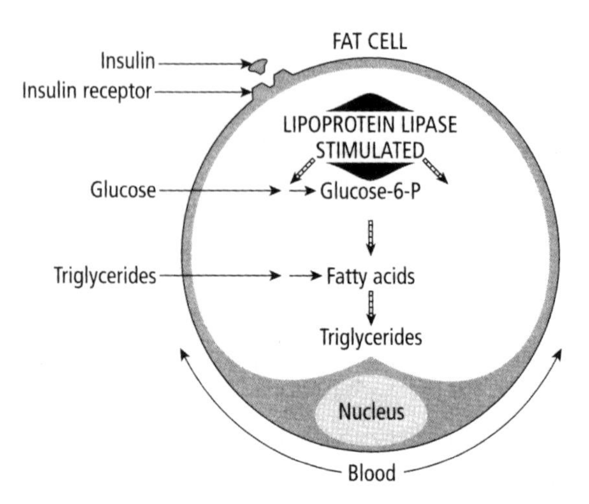

Figure-1: Excess insulin boosts fat storage by increasing LPL

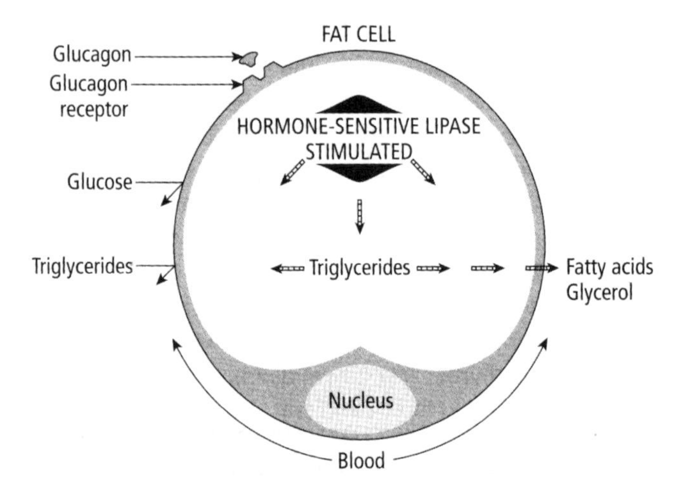

Figure-2: Glucagon stimulates HSL, which encourages fat burning

Hormonally (and, of course, enzymatically) speaking, when it comes to dietary intake, you really are only as good as your last meal. Many unbalanced diets disrupt the natural metabolic balance of storage and release activity that occurs in the fat cells. This imbalance can lead to becoming over-fat or obese. The key to an effective metabolism is to maintain a healthy balance between insulin and glucagon.

Human Growth Hormone: A True Messenger of Metabolism

Remember the childhood I described earlier, in which you were so exhausted from exercise by the end of the day that you were fast asleep before your head hit the pillow at night? And remember how you would wake up in the morning ready to start all over again – and without your morning java fix? That's because a body runs at peak metabolic power only when it is given down-time to build and repair. In other words, we need to sleep to enhance our anabolic metabolism. While you may be under the impression that your superhuman body only needs two or three hours of sleep per night, the truth is that denying yourself adequate rest robs you of vital metabolic resources, which in turn accelerates catabolic metabolism and bodyfat accumulation.

Sleep, Glorious Sleep

A population-based study appearing in the January 2005 edition of the *Archives of Internal Medicine* shows such a link. Researchers had 1,001 patients from four primary care medical practices fill out questionnaires concerning demographics, medical problems, sleep habits and sleep disorders. The conclusions were that insufficient sleep is associated with decreased metabolism, leading to overweight and obesity.[13]

Researchers from Stanford University School of Medicine have discovered that the alteration of the two key hormones *ghrelin* and *leptin* may be a key reason as to why tired bodies experience dysfunctional metabolisms and accumulated body fat. Ghrelin is one of the primary metabolic hormones responsible for enhanced hunger signals, while leptin works to suppress appetite. The ground-breaking study using the "real life" conditions of over 1,000 volunteers appears in the December 2004 issue of *Public Library of Science*. It shows that subjects with an average of five hours of sleep per night experienced a 15 percent increase in ghrelin and a 15 percent decrease in leptin compared to those who slept approximately eight hours on average.[14] These hormonal changes are enough to lead to enhanced hunger and fat gain. Other research, from the University of Chicago, indicates that healthy, sleep-deprived subjects reported a 24 percent increase in appetite – especially for sweet, starchy and salty foods.[15]

If you've ever found yourself channel-surfing well past the time you should have been in bed, you have no doubt come across an infomercial or two on an incredible elixir of youth called human growth hormone or HGH for short. If you also found yourself dialing the 1-800 number on the screen in hopes of obtaining the miracle product that all those incredibly healthy people on the screen were raving about, you would have been better off keeping your credit card in your wallet.

HGH is a protein-based hormone that is released by a tiny structure in the centre of your brain, called the pituitary gland. HGH is produced in the anterior region of the gland. Although one of its main roles in the body is to regulate growth, especially at puberty, it is also responsible for keeping your metabolism functioning at peak capacity by maintaining and – with the aid of testosterone – increasing lean body mass to make sure your body fat is used for fuel instead of storage.[16]

"Funny enough, if you had just been sleeping instead of watching late-night television, you probably would have made enough HGH to present your own testimonial to the night owls of the world. Sleep on that!"

The majority of our hormones are controlled by 24 hour cycles called circadian rhythms.[17] When it comes to HGH, two-thirds of this hormone is produced while we are in our deepest phase of sleep (Stages III and IV).[18] HGH levels are also regulated by two opposing hormones produced in another section of the brain called the hypothalamus. Growth hormone–releasing hormone (GHRH) is responsible for increasing the amount of HGH in the bloodstream and somatostatin is responsible for decreasing or halting HGH production.[19] In addition to a good night's rest, intense exercise –

particularly resistance training – is the next best way to maintain peak levels of this hormone.[20]

The most important factors for achieving optimal HGH levels are:

- going to bed as early as possible, with sound sleep from 11 p.m. to 2 a.m.;
- making sure your body has enough quality protein;
- performing regular, strenuous exercise;
- losing excess body fat;
- controlling excess stress;
- remaining free of disease.

Funny enough, if you had just been sleeping instead of watching late-night television, you probably would have made enough HGH to present your own testimonial to the night owls of the world. Sleep on that!

The False Hormones: Xenoestrogens

Over the last several years, researchers have focused on the role our sex hormones play in the way we look, feel and perform – not to how much belly fat we accumulate with each advancing year. At the center of this debate is the group of hormones known as the estrogens.

When many of us think of estrogen, we usually consider the three natural estrogens the human body produces – estradiol, estrone and estriol. Estradiol is the most powerful and high levels are believed to be a major cause of cancer. Estrone and estriol are manufactured (metabolites) from estradiol. However, they are not nearly as powerful mediators of cellular growth as their predecessor. Estriol is the weakest estrogen in the body. There are 11 other estrogens produced by the human body, all of which come from estradiol. When researchers think of estrogen, however, they not only think of the natural estrogens produced by the body, but of many other things that resemble estrogen once inside the body. As unbelievable as it might seem, foreign substances can enter the body and mimic our hormones, causing all sorts of unscheduled message-sending. When these chemicals resemble estrogen, they are referred to as estrogen-mimics or xenoestrogens. ("Xeno" simply means false.)

Xenoestrogens find their way into our bodies from pesticides, plastics, food supplies and our environment. When a xenoestrogen disrupts our natural hormonal system, it is referred to as an endocrine disruptor. Because these endocrine disrupters are fat soluble, they often become lodged within our fat cells and are very difficult to get rid of.[21] When it comes to your metabolism, endocrine disrupters are known to cause a disturbance in the way your body metabolizes important biochemicals.[22] It is also believed that endocrine disrupters create an enhanced environment for our bodies to store fat and make it extremely difficult to lose it.[23]

Most alarming, endocrine disrupters greatly increase our risk of cancer – especially breast cancer in women and prostrate cancer in men. In fact, research presented in the May 2004 issue of the *Journal of Applied Toxicology* indicates that endocrine disrupters commonly found in many body care and cosmetic products may prove to be more dangerous than once believed. Ingredients such as p-hydroxybenzoic acid esters or parabens have been shown recently to be highly estrogenic. These xenoestrogens have also been detected in human breast tumor tissue, indicating that they are definitely absorbed through the skin.[24]

How Do Xenoestrogens Cause Problems?

In order for hormones to carry out their biologically intended functions, they must first activate special proteins called receptors. These receptors are like switches that can only be activated by specially shaped fingers. In the case of natural estrogens, their unique shapes turn on and activate estrogen receptor switches.

According to research published in the *Journal of Endocrinology* in 2002, endocrine disrupters work in a similar manner, although instead of activating the same receptors natural estrogens activate, they instead activate a newly discovered group of receptors called estrogen-receptor-related receptors.[25,26] Estrogen-receptor-related receptors are only activated by xenoestrogens, and their biological messages are often quite disruptive to healthy metabolism.[27]

The Hormonal Circle of Life

Metabolic research is now showing that our overall hormone metabolism changes with age. The end result is diminished youth, energy, vitality, muscular size and strength, and an ever-increasing waistline, not to mention an enhanced risk of heart disease and cancer.

As Michael A. Zeligs, MD, a physician and nutritional expert from Boulder, Colorado, states, "Slower hormone metabolism in mid-life can mean higher than normal levels of estrogen and a deficiency in its beneficial 2-hydroxy metabolites."[28] These 2-hydroxy metabolites of estrogen are often called the "good estrogens," as they are able to support healthy hormone balances, protect against certain forms of cancer, especially cancers of the breast, uterine and prostate, as well as helping to mobilize stored fat in both women and men. It is the overproduction of bad estrogens (the 16-hydroxy variety) that negatively affect metabolism and contribute to body fat stores as well as other health challenges.

"Evidence of hormone imbalance for women includes PMS, difficult perimenopause and menopause. Men might notice muscle wasting, depression and prostate problems. Both sexes can experience loss of libido and weight gain, especially in the abdominal area."

This hormone imbalance is also exacerbated by the multitude of endocrine disrupters we come in contact with on a daily basis. Evidence of hormone imbalance for women includes PMS, difficult perimenopause and menopause. Men might notice muscle wasting, depression and prostate problems. Both sexes can experience loss of libido and weight gain, especially in the abdominal area.[29,30,31] (See Part Two, Her Health and His Health) Remember, it is possible to diminish the frustrating symptoms of hormone imbalance – or avoid them altogether – by giving your body what it needs for vital metabolic power. And be sure to avoid endocrine disruptors by eating organic foods and using non-toxic body care products.

Men and Women – The Metabolic Difference

One of the frustrations many women express to me is that their fat cells seem to expand so much more easily, and yet deflate much more slowly in comparison to their male counterparts. If you've also wondered why men and women seem to be so metabolically different in this regard, allow me to shed some light on this perplexing and often hair-pulling phenomenon. But before I begin, keep in mind that I am merely the messenger here.

Women carry at least 8% more bodyfat than men on average. That's equal to an extra 120,000 calories in storage![32] This is partly due to women's metabolic disadvantage. Maintaining muscle mass and function through age can offer you a huge metabolic advantage when it comes to burning bodyfat. The truth is, men have a lot more muscle. In fact, men usually sport about 40 pounds more, and due to this variance in metabolically-active tissue, men can burn up to 30% more calories than women during exercise as well as sitting on their butts doing absolutely nothing! So, while you're sweating your buns off on that new elliptical thing-a-ma-jiggy, that certain man in your life is burning the same amount of calories playing with his remote – well, almost as many.

Now, please don't think that I am placing *all* the blame on the difference between a few pounds of muscle. There are also differences in the quantities of two very important hormones – estrogen and testosterone. Yes, women's bodies contain testosterone, just as men's contain some estrogen. These hormones dictate to a large extent how much muscle you currently have, not to mention how well that muscle is behaving.[33] Aside from their role in regulating sexual and reproductive function, these hormones also help control your metabolism.

Ask any bodybuilder and they will tell you that it is next to impossible to build a lot of muscle mass without heaps of testosterone. For both men and women, adequate testosterone levels are imperative to maintaining and reactivating metabolism. Testosterone also helps control blood sugar, which means that it helps minimize insulin levels – and therefore fat manufacture and storage.[34] A woman produces about a tenth of the testosterone a man produces, while at the same time producing a lot more estrogen. This equation shifts quite a bit, however, as men age.[35]

It is important to note that not all testosterone is created equal. It is only the free, biologically active (or "unbound") testosterone that dictates how testosterone reacts in the body. Inactive testosterone actually becomes bound or tied to a specific protein known as sex-hormone-binding globulin (SHBG), and is therefore prevented from doing much. Unfortunately, as men and women age, it's the active form of testosterone they lose.[36] This is also one of the main reasons we lose muscle and gain body fat.

Research indicates that restoring testosterone to optimum levels can reverse many of the so-called signs of aging. Along with an increase in muscle mass and a corresponding decrease in body fat, restored testosterone levels can also improve brain function, cardiovascular function, bone strength, and overall mood in both men and women.[37,38]

So, what is the best way to restore testosterone? The first is to maintain a well-balanced diet with optimal levels of protein, fat and fibre.[39,40] (I'll go into more detail about a metabolism-boosting menu in the coming chapters.) Researchers from the University of Massachusetts Medical School have discovered that elderly men with low dietary protein intakes have elevated SHBG levels and lower amounts of active testosterone.[41] Adequate protein intake becomes one of the keys to boosting active testosterone levels.

Pump Up the Volume

Proper exercise strategies will also help you pump up your testosterone levels. But contrary to popular belief, it doesn't happen with cardio activity. Only weight-resistant exercise has been shown to raise testosterone levels effectively.[42] Don't misunderstand me; I'm not saying that cardio is ineffective. In fact, it's extremely important for your cardiovascular system to get a work-out. What I'm saying is that cardio exercises are not as important in terms of restoring your testosterone. (See Chapter 9, Exercising Your Metabolism.)

And Lighten Up on the Stress

When it comes to the stress response we have inherited from our ancient ancestors, perception is the same as reality. In other words, your body reacts to an actual or an imagined stress exactly same way – by flooding your body with stress hormones. And studies show that levels of our much desired anabolic testosterone can decrease dramatically (sometimes by as much as 50%) in times of stress.[43] The problem is that stress hormones are produced along the same biochemical pathway as testosterone. Stress hormones tend to gobble up most of the raw materials needed to manufacture testosterone, causing testosterone levels – and you – to suffer as a result. We also know that nothing helps to alleviate stress better than a good work-out, so the next time a situation makes you feel like you want to throw a heavy object, channel that energy into a testosterone-building, weight-bearing work-out!

Metabolic Aging

As we get older, we tend to notice a loss of muscle mass, strength, bone mass and our once vital energy potential, all the while experiencing body fat increases, wrinkling of the skin, various mood changes and a declining immunity – oh joy! New scientific research is shedding light on why we age, and how we may be able to slow the process.

There are numerous theories explaining the aging process, however, most of these theories eventually come to the same conclusion; aging is a process of accumulated damage (catabolism) to our genetic blueprint, our DNA. If the DNA in one of your cells were to unwind to its fullest potential, it would stretch over one meter in length, which is why it is tightly fitted into compact units or DNA packages called chromosomes.

Over the last few years, longevity researchers have focused their attention to the caps on the ends of these chromosomes called telomeres. Every time our cells divide, the telomere gets a little shorter until cellular division comes to a halt. Once a cell has run its full potential of cellular divisions, it is referred to as a senescent or old cell.

Since aging is a process of the body becoming unable to repair accumulated damage, longevity researchers believe that cellular senescence may be one of the key causes of aging, with telomere damage leading the way. Research presented in a 2001 edition of the *Journal of the American Geriatric Society* indicated that telomere shortening is directly associated with accumulated DNA damage and cellular aging.[44]

According to Dr. Judith Campisi, one of the pioneers in the research of cellular senescence from the Lawrence Berkeley National Laboratory in California, senescent cells don't die once they stop dividing, but instead hang around emitting harmful proteins to neighboring cells, eventually leading to cellular malfunction. Research presented in the *Journal of Experimental Gerontology* in 2001 showed that cells with dysfunctional telomeres could contribute to cancer and aging, as telomeres are essential for maintaining the integrity and stability of our genes.[45]

It is also widely known that as we age, we often become more insulin resistant. Researchers have always known that insulin resistance and accumulated body fat can predispose people to cardiovascular disease and a shorter lifespan. The mechanisms behind this have not been as easily understood – perhaps until now. Research published in the May 2005 issue of the journal *Circulation*, showed that increased insulin resistance along with a higher body fat content results in greater telomere shortening over time. This study is the first to show tangible evidence that insulin resistance and excess body fat leads to premature aging.[46] But it's only one example.

As you read through the material presented in *Awaken Your Metabolism*, you'll learn that, along with hormones, diet, exercise and rest, supplemental nutrients impact metabolism and aging. Most important, as you consider these factors, remember that every cell in your body has its own metabolism. If treated right – that is, provided with the right fuels and lifestyle choices – that metabolism will allow your trillions of cells to work in harmony. The outcome, simply enough, will be a disease-free life, lived with vitality and abundant energy.

CHAPTER 4

FEEDING YOUR METABOLISM

Okay, so we agree that our goal is to restore and maximize our fullest metabolic energy potential. In order to accomplish this, we must work to revive our cellular energy factories to first boost metabolism at the cellular level and then, by extension, throughout all of our body systems. So the question is, what do our cells need to accomplish this? Thankfully, the answer is quite simple: nutritious food (including effective nutrient supplementation) and sufficient exercise, as well as adequate rest and recuperation in the form of sleep.

What's that? Did I hear you groan? Are you now concerned that this book is going to describe a nutritional regimen that is just way too stringent coupled with an exercise program that would fell an Olympic athlete? Relax. I would never suggest such a thing. Life is too short to watch every morsel of food you put in your mouth or to spend your entire day in a gym. Besides, who would want to? Instead, I am going to introduce you to a new approach to eating and later to moving your body properly through exercise. And this strategy doesn't require you to count a single calorie or even join a gym – if that's your choice!

Contrary to popular belief, the way to enhance metabolism is not to starve your body but instead to feed it. In essence, your metabolism actually needs food in order to thrive. Starving it will actually cause the opposite effect, slowing it down to a snail's pace! One of the reasons our metabolism tends to slow down with age is because older individuals tend to eat less than younger ones. Researchers from the University of Colorado were able to back up this assumption when they compared two groups of older and younger inactive men. The metabolic rates (measured by BMR) of the older subjects indicated that they burned a maximum of 1,632 calories per day, compared to a maximum of 1,848 calories per day in the younger subjects. But when both groups performed the same amount of exercise and/or consumed the same number of calories, their metabolic rates were almost identical.[47] So, rather than offering you some complicated patent-pending diet plan, I'm happy to report that the first step to achieving optimum metabolism is to give your body the right foods in the right quantities and at the right times.

Paleolithic Nutrition

In terms of our evolutionary history, we haven't been on this planet very long, nor have our bodies changed much in that time. Humans have been around for nearly 2.5 million years, but according to leading geneticists, we share 99% of the same genetic structure as our prehistoric brothers and sisters, dating back at least 40,000 years.[48] This is way, *way* before the agricultural revolution of approximately 12,000 years ago – a mere blip in terms of evolutionary time (one half of 1 percent of our evolutionary history).[49] In other words, our bodies thrive on a basic and primitive diet, not a diet that is contaminated by the modern principles of food processing. Processed food, of course, includes anything in a can, box or freezer-pack. Sociologist Alvin Toffler in his bestselling book, *Creating a New Civilization: The Politics of the Third Wave* (Turner Publishing, 1995) said, "Something on the order of 1,000 new products are introduced into America's supermarkets every month."[50] How can we ever expect our biochemistry to adjust to this bombardment of newfangled foods? We can't. The truth is, our bodies operate best when we feed them the same foods our ancestors ate for thousands of years. And that doesn't include a mainstay of potato chips, white bread and pasta. Read on.

Learn from our Ancestors

Centuries ago, our hunter-gatherer ancestors functioned best on a diet that consisted primarily of lean sources of protein, supplemented with whole fruits, vegetables, nuts and seeds. In fact, research presented in the best-selling book, *Dangerous Grains*, notes that humans were using tools and hunting meat as long as 2.5 million years ago. If you want to keep your metabolism churning at high speed and fend off disease, you would be very wise to adapt the same way of eating.[51] According to Dr. Eaton, the less we eat like our ancestors, the more we put ourselves at risk for diabetes, heart disease, arthritis, and cancer. And, of course, we cannot forget the additional risk of obesity.

We are not designed to consume highly processed foods, laden with sugar, artificial sugar, or white flour and cornmeal based products. We're just not built for it. I don't care how good Grandma's buttermilk biscuits and jello salad taste; these foods should not be the dominant source of your daily caloric intake! Would you put cheap fuel in your new sports car? Of course not!

In the next few chapters, I'll describe the metabolism-boosting foods we need on the daily menu. You'll see quickly enough that I don't want to starve you. In fact, as mentioned above, reducing your caloric intake to a level too low is a recipe for metabolic disaster.

Feast, Famine and Your Fat Cells

We are genetically engineered to thrive in a "feast or famine" environment.[52] As a means of survival, our fat cells evolved to expand without a mechanism to turn them off. The potential of your fat cells to increase is enormous – as if many of us didn't already know that! Our 30 billion fat cells can expand up to 1,000 times their regular size.[53] Don't get mad at them. They are just doing their job and they are the reason you are here today. They store all that fat as fuel so there will be enough food readily available the next time your body goes into starvation mode, usually during a weight-loss diet. In a world where the necessity for the feast and famine principle has been replaced with multitudes of convenience and fast food establishments, and where we are becoming increasingly sedentary, it is no mystery why we are getting fatter.

Ironically, in the midst of such abundance and growing obesity, many of us are still starving to death on full stomachs and digging our graves with our spoons and forks. Instead of dying from lack of food, we are killing ourselves, literally, on French fries and excess sugar. Despite the emergence of a drive-through lifestyle, we are still genetically wired to consume the same foods our ancestors ate over 30,000 years ago.[54] And by constantly making unhealthy lifestyle choices, we end up damaging the hormonal and biochemical balance of our bodies. Excess stress, lack of sleep and little or no exercise contributes to our metabolic decline. Before we know it, we become one more statistic in the couch potato plague.

Then, when we can't stand our lifestyle a moment longer, we rush head-long into the newest, flashiest diet that our friends are trying. And how successful has that been in the past?

The Problem with Diets

As you read this, over 40 percent of women, and 28 percent of men are starting or finishing another fad diet.[55] Although North Americans spend over 33 billion dollars per year on the diet industry, (yep, that's billion with a "b") we are still getting fatter.[56] Guess what? In case you haven't noticed by the present state of our society, the majority of diets don't work. How many times have you decided to bravely embark on the challenge of strict calorie counting, abstaining from fat, and all the other evils in the diet world? You were determined to put all your weight "issues" behind you, this time for good! Really!

After all that sacrifice, you have successfully dropped two sizes and your bathroom scale is finally moving in the right direction (actually, to the left is more like it!) As you stand in front of that full-length mirror, however, there isn't a lean, mean machine in front of you, but instead a smaller fat person. In fact, to the naked eye, your body fat looks pretty much the same as when you started your diet.

So, what the heck happened? Well, 99 percent of diets fail miserably. We are just not designed to go on fad diets. In fact, they go against thousands of years of our evolutionary history. The American Medical Association will actually tell you that the gene pool has remained virtually unchanged throughout obesity's rise.[57] And, you'll remember that our genetic makeup is still programmed for a feast or famine environment. Although we no longer have our ancestors' lifestyles (thank goodness, you might say!), we still wear their genes of thousands of years ago. There were no diet books back then, or burger havens or dessert bars for that matter, to contribute to their weight problems.

In feast and famine mode, the body will do anything in its power to make sure you survive. It does not recognize that there is a multitude of eating establishments all around you. It also does not recognize the difference between true starvation, or an unhealthy restriction of essential nutrients through dieting.

"Another major reason most diets fail is because the people going on them rarely – if ever – consider what is being lost as the numbers on the scale scroll backwards. Research shows that for every pound of fat you lose on an unbalanced diet, you can also take with it nearly one pound of muscle!"

When you diet, you fight your genetic makeup. The body goes into panic: "I'm being starved again, but this time it's worse!" So after you finish your diet, not only do lost pounds reappear, extra weight is piled on to make sure that you will survive the next famine – also known as the next unhealthy fad diet. You must realize that fighting your genetic make up only leads to disease and a sluggish metabolism.

Another major reason most diets fail is because the people going on them rarely – if ever – consider what is being lost as the numbers on the scale scroll backwards. Research shows that for every pound of fat you lose on an unbalanced diet, you can also take with it nearly one pound of muscle![58] With the loss of muscle comes a declining metabolism and, you guessed it, more fat. This makes sense since muscle is the major fat-burning engine in the body.

Loss of muscle tissue is the primary reason for rebound weight gain. We need to do everything in our power to maintain this invaluable metabolic substance. Losing even *one* ounce of our precious muscle lowers the body's metabolic capacity. To maintain an optimal metabolism, you need muscle – period! One pound of muscle can burn up to 50 calories a day. The leaner you are, the more calories you will burn in a 24-hour period.[59]

Boosting the metabolism is not about dieting, but it is about eating a healthy diet. The next several chapters will explain the right foods for creating and sustaining your metabolic fire.

CHAPTER 5

DIETARY PROTEINS

While the goal in this book is to show you how to boost your cellular metabolic engines in order to live a life of increased energy and vitality, one of the very obvious first indications of optimum metabolism is ongoing fat-loss. This is a great thing, not only because of the health consequences of carrying too much weight, but also because, as I discussed in the first chapter, increasing our percentage of lean muscle has the power to boost our longevity as well. Oh, yeah, and you'll look better, too.

One of the best ways to kick-start your metabolism is with dietary protein. Although we no longer have to club our food over the head and drag it home as our ancestors did, we still need to feed our cells with high quality lean protein. Remember high school science? "Protein provides the building blocks for our body." We should have been paying more attention, instead of twirling our pencils and staring into thin air. Our once tiny butts would be thanking us. Eating proper protein will rev up your metabolic engines![60]

Dietary proteins are absolutely essential for survival (let alone for fat-burning success!) In case you missed science class the day your teacher discussed protein, it is actually the amino acids in protein that are the building blocks for our organs, muscle cells, skeletal system, transport proteins, enzymes, immune system, hormones, chemical messengers, etc.

Dietary protein is the substance that stimulates the fat-burning hormone glucagon, as well as building metabolically active muscle.[61,62] Another incredible thing about protein lies in its ability to promote a process called thermogenesis – a fancy word for heat production.[63]

Thermogenesis acts like a metabolic furnace that creates extra heat for our bodies by incinerating your body fat reserves. This means that we can literally burn fat!

According to studies of prehistoric populations, our ancestors who lived on a Paleolithic diet were tall, lean, muscular, had very little tooth decay and not a lot of disease.[64] They were healthier, it appears, than the grain-fed, carbohydrate loving soon-to-be-junkies of the early agricultural revolution. They were the superstars of evolutionary eating.

And they ate a lot of protein! The protein they consumed was a lot healthier than the majority of protein we consume today. This was mostly due to the health of the animals they hunted, not to mention the fact that plant foods were free of chemical pesticides, fungicides and herbicides. Protein helps you elevate your resting metabolism throughout the day.[65] If you eat enough high quality lean protein, you can even burn calories while you sleep.

When it comes to protein, however, it is not as important *how much* you eat, but rather how much of the protein is *available to the body* after ingestion.[66] Five processed pepperoni sticks will not do the same trick as some better, healthier protein choices. Protein requires a high biological value in order to perform its structural duties (muscle, bones, skin and organs) and functional duties (immunity, hormones, enzymes and peptides) at peak capacity. After all, you really are only as strong as your body's ability to keep up with the daily wear and tear on your trillions of cells. This is where protein shines. High biologically-valued protein choices – consumed in the proper proportions and at the right intervals throughout the day – allow your body to remain in an anabolic state, which as you now know, means that you preserve a youthful metabolism by maintaining and activating your lean body mass.[67]

The higher the quality of protein you consume, the faster you will see yourself looking at a leaner physique in the mirror and feeling the energy that was once so familiar to you. High quality protein choices like breast meat from free-run chickens and turkeys, fish, free-run eggs and lean meats like wild game and tenderloin cuts will help you stoke the metabolic fire better than a low protein, high carbohydrate diet.[68] Some sources of dairy are also good protein choices, especially if they come from organic goat milk. High-quality whey protein isolates (especially ones high in alphalactalbumin) are also an excellent source, and I will explain their benefits later.

Eating protein in the morning will start your day off right! MIT researchers analyzed a group of people who had eaten either a high-protein or high-carbohydrate breakfast. Both groups started out with similar levels of the sleep-inducing compound called tryptophan in their blood. Two hours later, those who ate the high-carb breakfast had more than four times the level of tryptophan in their systems compared to the high-protein group, and were thus more likely to feel tired and sluggish. As you will see a little later, too many carbohydrates can easily put your metabolism on "slug" mode, making you a fatter, energy-deficient version of your former self.

Obtaining a "lean-mean-metabolic machine" is not only possible, it is highly likely with the right dietary choices. And protein is always leading the pack – or the six-pack if you prefer!

PROTEIN AT-A-GLANCE:

Protein helps:

Stabilize blood sugar
Promote cell growth and repair
Hormone production
Enzyme production (digestive and metabolic)
Neurotransmitter production
Cell metabolism
Body fluid balancing
Maintain your immune system

Excellent Sources of Dietary Protein

It is important to choose organic whenever possible or free range and medication free if organic is not available.

MEAT

All game meats (especially bison, venison and elk)
Tenderloin
Liver (organic baby beef)
Chicken breast (skinless)
Turkey breast (skinless)
Cornish hen
Egg (white or whole)
Whey protein isolates (cold-processed, high alpha)

SEAFOOD

Salmon (wild)
Bluefish
Snapper
Haddock
Trout
Tuna steak
Cod
Clams
Shrimp
Bass
Sardines (in water)
Sole
Swordfish
Mackerel
Halibut
Scallops
Crab
Lobster

DAIRY

Cheese (low fat percentage only)
Cottage cheese
Organic Milk (preferably goat—1% to whole)
Yogurt (only organic plain)

VEGETARIAN

Hemp
Soy (only organic fermented)

CHAPTER 6

DIETARY FATS

Have you ever been called a fat head? If you have, take it as a compliment. The dry weight of your brain is 60 percent fat.[69] Fat also exists in the membrane that covers the trillions of cells in your body, and the raw structure of our eyes, ears, sex and adrenal glands. The point is, not all fat is created equal and we need lots of the right fats to maintain a healthy metabolism. In fact, the only fat you should ever be afraid of is the excess body fat that is presently stripping you of your life potential. The key to good health is to know how to choose your fats and to supply the right amount of these fats to every one of your trillions of cells.

The fact remains that if you continue to consume the wrong types of dietary fat (as well as high glycemic carbohydrates – more on this in the next chapter), you could very well be setting yourself up for a declining metabolism.[70] This is one of the main reasons the majority of North Americans are dragging themselves around in a perpetual state of lethargy. The good news is that you can easily correct this metabolic deficit by consuming a proper balance of the right type of dietary fats. In other words, proper fatty acid intake and supplementation can help you experience continual fat burning, as well as increased energy levels.[71]

Remember the low fat craze that started around twenty years ago? In case you haven't noticed, it didn't exactly solve the obesity epidemic we are now facing. In fact, over this time, we have experienced a whopping 60 percent increase in the incidence of obesity. Research shows that as long as we keep our consumption of good fats to a moderate level (somewhere between 20 and 30 percent of our total daily calories) we should not get fat.[72]

So many of us have tried to cut the fat only to find, to our astonishment, we kept getting fatter! There is no malicious universal force at work here. The fact of the matter is when you restrict your consumption of fat, you deprive your body of *essential fatty acids*. Yes, you read that correctly. Some fats are essential in your diet. Every single cell in the body needs fatty acids. We should not be restricting them, but instead consuming them in sufficient quantities to maintain optimum health. When you cut fat to the extreme, not only do you rob yourself of essential fatty acids, you end up making up the caloric deficit by consuming a large amount of carbohydrates, often in the form of the low-fat/no-fat foods that have popped up over the last 20 years.

The bottom line is, your body literally thrives on essential fatty acids – also referred to as omega-6 and omega-3 fats. Essential fats keep your cells happy and encourage them to burn more bodyfat by maintaining peak metabolic activity. In other words, proper fatty acid supplementation can help you experience continual fat burning as well as increased energy levels.

For instance, the omega-3 fatty acid EPA found preformed in fish oils has been shown to lower blood levels of fats by increasing the transport of fatty acids into the cells for fat burning (thermogenesis). Many researchers believe that well over half the North American population is deficient in these omega-3 fatty acids. Perhaps our current obesity epidemic is at least in part due to this deficiency.

Back to the Basics

The diet of early man consisted of mainly wild game, fish, nuts, seeds, roots and shoots. Natural whole foods contain healthy fats as part of their natural composition, with an ideal balance of saturated, monounsaturated, and polyunsaturated fats. Eating the wrong fats (such as man-made trans-fats found in numerous processed foods or too many saturated fats from animal products) will cause your metabolism to become sluggish and your waistline to grow like there's no tomorrow.[73] If you care about the future state of your metabolism (not to mention your 30 billion fat cells), you need to avoid these fats like the plague.

Polyunsaturated Fatty Acids (PUFA)

OMEGA-6 FATS (linoleic acid) are found primarily in polyunsaturated oils like corn, cottonseed, safflower, soybeans, sunflower, walnuts, and wheat germ. Your body required between three and six percent of your total daily calories from these fats.

OMEGA-3 FATS (alpha-linolenic acid) are found in large amounts in flaxseed and Perilla oils, and in lesser amounts in canola and hemp seed. The main benefits derived from omega-3 fats relate to their ability to convert into two health-promoting fats – eicosapentaenoic acid (EPA) and docosahexaenoic acid (DHA) – which are found preformed in cold-water fish.

Monounsaturated fats (MUFA) are another group of fats important to your overall health and metabolism. In fact, recent research published in the April 30 issue of the *British Medical Journal* indicates that the Mediterranean diet – which is high in MUFA – may not only be one of the healthiest diets around, it may actually be able to extend human lifespan.[74] MUFA are found in olive, avocado, canola and high-oleic safflower and sunflower oils. MUFA help to balance your metabolism. However, your overall metabolic success lies within a healthy balance of the omega-6 and omega-3 polyunsaturated essential fatty acids in the chart above.

Though experts cannot agree on the exact ratio of omega-6 to omega-3 fats we should be consuming, most agree that it lies somewhere between 1:1 and 4:1 (four parts omega-6 to one part omega-3). The problem today is that the ratio has shifted much too far in the omega-6 direction: modern estimates place our adult dietary ratio anywhere between 20:1 and 50:1.[75] In essence, modern diets consist of far too much omega-6, which contributes to many modern health challenges. That's why it's so important to boost your intake of omega-3 fats.

If we do not eat an adequate supply of omega 3 fats, our metabolism will suffer and fat burning will come to a complete stop. Can you imagine fat actually helping you burn fat? These cooperative little essential fats work together to increase the oxygen utilized by the cells to produce energy. This is wonderful because the more oxygen we transport to our cells, the better our metabolisms function, and the more fat we burn! In fact, if you continue to consume the wrong types of dietary fat, you could very well be setting yourself up for a low energy status and continual fat storage by degrading your metabolism. This is one of the main reasons the majority of North Americans are dragging themselves around in a perpetual state of lethargy.

Although fats are essential to the diet, fats should never exceed 30 percent of our daily dietary caloric intake.

Fats At-A-Glance

Healthy Fats help:
Balance blood sugar
Provide raw materials for hormones and other messengers
Create fuel for long-term energy
Strengthen cell walls and mucus membranes

Excellent Sources of Fats:

Flax oil
Hemp oil
Coconut oil (organic/extra virgin DME) – great for cooking
Olive oil (organic/extra virgin) – good for cooking at lower temperatures
All nuts and seeds (especially almonds, walnuts and pumpkin)
Almond butter
Cashew butter
Avocado
Medium chain triglycerides (coconut source)

CHAPTER 7

DIETARY CARBOHYDRATES

Are you of a member of "the Carboholics Association of North America?" If not, you are one of the lucky ones. The vast majority of North Americans are now on the Board of Directors. For the most part, the majority of us are carbohydrate addicts – consuming them as our main source of food, with dietary proteins as a supplement.[76] According to evolutionary expert Dr. Boyd Eaton, however, the early human diet – the one we should be eating – consisted of at *least* 30 percent protein. We have been consuming carbohydrates as our main dietary source for the last 12,000 years, and the over-processed kind for less than 100. Other experts, such as Loren Cordain Ph.D., from Colorado State University in Fort Collins, Colorado agree that our bodies have not yet made the genetic alterations necessary to thrive on such a high carb diet.[77] In fact, it will probably be at least another 10,000 years until our genes catch up to our present way of eating. I don't think you and I have enough time to wait!

Consider this: the average North American citizen consumes a diet of 50% carbohydrates (most of them processed), 34% fats (and not the right kinds, as I discussed earlier), and only 15% protein (most of it not lean cuts, and much of it highly processed).[78] Is it a good formula? Our lack of energy, ever-expanding waistbands and increase in disease over the last 20 years should be evidence enough to convince the average citizen of a need to make nutritious dietary changes. But, instead of improving on this inadequate dietary profile, we continue to consume higher percentages of the wrong foods, including high glycemic carbohydrates. The glycemic index (GI) measures the rise in circulating blood sugar triggered by a carbohydrate. A food with a low GI response will cause a small rise, while a high GI food will cause a sharp spike. We need only consider the increase in type 2 diabetes in recent years to see how this menu plan impacts blood sugar.

The Story of Sugar

Whether you consume 1/4 cup of refined sugar or 1/4 cup of a baked potato, in the end they will both be converted to 1/4 cup of sugar in your body. No, this is not a riddle, or a Zen meditation like "what is the sound of one hand clapping?" It is a scientific fact. Does this mean a baked potato is just as bad for you as refined sugar? No, potatoes are a source of potassium, iron, fibre and vitamins, while sugar is packed with empty calories.

But since your body can only deal with so much sugar at one time, if you consume too much, it can easily wreak havoc on your delicate hormonal balance. Highly-processed, refined carbohydrates can raise your insulin levels through the roof, very quickly. If you want to maintain an optimal metabolism, one that burns fat around the clock, elevated insulin levels can be your worst enemy. If we have too much insulin in our systems, the body cannot access bodyfat as a fuel source. The majority of over-fat people have high resting insulin levels, not only in their blood but also in their cells. This, of course, slows down the fat burning engines.

Highly processed carbs that raise blood sugar levels come from foods that are high on the glycemic index.[79] You will recall that carbohydrates are ranked by how fast they enter our bloodstream and cause a rise in insulin. Foods low on the glycemic index are your best choices. This does not mean, however, that you can eat as many of them as you wish, as balance is always the key.

It is important to note that just because a food is high on the glycemic index does not mean you need to boycott it altogether. You can combine a small amount of high-glycemic foods with a larger quantity of low-glycemic foods and easily balance out of the equation; this is referred to as the glycemic load of a meal.[80] For instance, whole-wheat pasta cooked *al dente* takes a lot longer to break down in the body than pasta that has been overcooked. In most cases, the longer you cook a food, the more quickly it will be assimilated into sugar in your body. This causes your insulin levels to spike, which is what we want to avoid at all costs. Lower glycemic loads have been shown to reduce the risk for obesity and many chronic diseases.[81]

The best way to maintain an effective blood sugar balance, and keep our insulin levels intact is to eat like our great, great ancestors did – Paleolithically speaking. We need to get our carbohydrates in complex form so that they take longer to digest – and therefore prevent a steep rise in blood sugar. Look for your complex carbohydrates in fruits and vegetables. And if you need to consume grains, make sure they come from sprouted whole grains (not white pasta, white rice or white flour!)

CARBOHYDRATES AT-A-GLANCE

Carbohydrates help:

Energy production
Thyroid conversion
Muscular repair
Maintain proper balance of insulin to glucagon

Excellent [low glycemic] Sources of Carbohydrates:

LEGUMES

Adzuki beans
Black-eyed peas
Garbanzo beans
Navy beans
Split peas
Kidney beans
Black beans
Butter beans
Lima beans
Pinto beans
Haricot beans

VEGETABLES

Artichokes
Asparagus
Bok Choy
Broccoli
Cabbage
Cauliflower
Celery
Dark Leaf Lettuces
Eggplant
Endive
Garlic
Green Beans
Mushroom (Portobello)
Onions
Radicchios
Red Peppers
Rhubarb
Spinach
Tomatoes
Yellow Peppers

FRUITS

Apples
Apricots
Avocados
Blackberries
Blueberries
Cherries
Cranberries
Grapefruit
Nectarines
Oranges
Peaches
Pears
Plums
Raspberries
Strawberries

GRAINS

Barley
Bran cereals
Bulgur
Cooked bran
Cooked oatmeal
Muesli cereals
Pasta, kamut and quinoa
Pasta, whole-grain, el dente
Pasta, whole meal, el dente
Porridge oats
Rice, red basmati whole-kernel
Whole-sprouted grains

CHAPTER 8

EATING PRINCIPLES

By now, you have a good understanding of how the dietary principles that guided us for thousands of years still apply to our biochemistry and govern our cellular metabolisms today. You are also aware of what foods to consume to take full advantage of your metabolic machinery. These foods include fresh, organic fruits and vegetables, nuts, seeds, whole grains and lean sources of protein. You are also aware of what foods hamper the system – at least the majority of the time. These include processed, high-glycemic foods such as white rice, white pasta and white bread.

Go Organic

Thousands of years ago we lived in a very different world. Beyond the obvious, the air we breathed was clean, the water we drank was crystal clear and the food we consumed was free of commercial contaminants. According to leading toxicologists, the annual production of synthetic chemicals weighs in at a yearly 600 billion pounds.[82] Where do you think the majority of these contaminants end up? You've got it – in the air we breathe, the water we drink and the food we eat. This information should help you appreciate why I believe organic is the best way to go whenever possible. After all, there is no need to further pollute your body.

- Organic foods are free from unnecessary chemicals, preservatives, pesticides, herbicides and fungicides.
- Organic foods come as close as possible to the way our ancestors ate.
- These foods supply an enhanced nutritive value to body structures. According to the *Journal of Applied Nutrition*, organic fruits, vegetables and grains contain approximately 90% more minerals than conventionally grown food.[83]

How to Eat

To ensure optimum metabolic success, be sure to:

- Eat five to six metabolically balanced nutrient-dense meals per day – ensuring that you eat every 2.5 to 3.5 hours to keep blood sugar stable. Nutrient-dense foods are those that are high in vitamins, minerals, essential fats, protein and fibre.
- Make sure that each of your meals is comprised of approximately 40% complex carbohydrates, 30% lean proteins and 30% healthy fats (This in no way needs to be an *exact* science.)
- Three of your daily meals are to be in solid form and two are to consist of liquid nutrition (i.e. you will prepare your own protein shakes.) This adds a highly convenient, yet powerful way to maintain anabolic metabolism and therefore enhance fat loss.
- These meals will consist of between 200 and 500 calories each depending on your lean body mass and your activity level. You don't want to feed the fat!
- Ensure that your last meal of the day is at least two hours prior to bedtime and try your best to make that meal no later than 8PM.

Water Wise

Water is the most important macro-nutrient for the body next to oxygen. It is essential to anabolism, the body's ability to rebuild and repair. Dehydration, on the other hand, is related to cellular breakdown. Biochemical water expert Dr. Batmanghelidij, MD, says at some point in our evolution, our body's signals for thirst and hunger went awry. Now, when we are dehydrated, we often misinterpret the signal for thirst as a signal for hunger.[84] Instead of reaching for a bag of chips when you think you are hungry, reach instead for the H_2O. Better yet, do as I do and add a specialized mineral sachet to your water container in order to alkalinize your water source, which allows your body to become better hydrated at the cellular level (for more information see X_2O in Appendix I.)

- It is important to remember that your body is 75% water, not 75% coffee, tea or pop. You must teach your body to accept and enjoy water as your beverage of choice.
- Most lean people also have the highest body-water content,

compared to obese people who have less than 50% body-water content. This percentage is roughly the same amount found in most of the hospitalized elderly.[85]

- Drink an eight-ounce glass of water approximately 20 minutes prior to consuming each meal.
- Never wait until you have a dry mouth or feel thirsty to drink water. By this time you are already in a state of dehydration.

Nutrients to Optimize Metabolism

While we all strive to eat the best possible foods, sometimes our bodies require a little help. We provide that help in the form of supplemental nutrients. These nutrients do not replace a healthy diet, they add to it.

Following are a few core products that I personally feel should be a part of just about anyone's metabolic enhancing program. I have found these three products to be more of a necessity rather than a supplement. Aside from the first three core products, you will find a mention of nutrients that have been documented to help enhance metabolism at the cellular level by supporting mitochondrial health and integrity.

Taking Protein the *High Alpha* Whey

When trying to maintain a highly functional metabolism, the last thing you ever want to do is break down too much protein. Any way you look at it, muscle cells enhance how many calories you burn over a 24-hour period.[86] Remember our earlier discussion of resting metabolic rate, or RMR. If you do not maintain muscle cell activity while losing weight, you will end up degrading your metabolic engines, and fat loss will come to a screeching halt!

Thankfully, new research presented in the prestigious *Journal of Nutrition* shows that by adding high quality proteins to the diet, people can maintain muscle mass while reducing body fat during weight loss.[87] In other words, it is easier to maintain a high metabolic rate if you consume the proper proteins. The study looked at protein foods that provided optimal levels of an essential amino acid called leucine. Leucine is one of a group of three amino acids (leucine, isoleucine and valine) called the branched chain amino acids, or BCAAs, that help regulate immunity, muscle growth and muscular activity, all of which are essential to maintaining an effective metabolic rate. BCAAs are essential amino acids, and in order to insure optimal health, they must be obtained from our diet on a regular basis. Optimal sources of BCAAs include beef, dairy products (especially whey), poultry, fish and eggs. Dr. Donald Layman, professor of nutrition at the University of Illinois, where the study was conducted, said, "Study participants following the moderately high protein plan were twice as successful in maintaining lean muscle mass."

Not only are BCAAs beneficial in maintaining a high metabolic rate, they have also been shown to aid in maintaining a powerful immune system.[88] It is well documented that long bouts of intense exercise often lead to immune suppression. In fact, prolonged intense exercise, such as long distance running (marathons and triathlons), have been shown through research to increase the frequency and severity of infections. This is one of the main reasons long distance runners and triathletes often experience cold or flu-like symptoms soon after a race.[89]

"Study participants following the moderately high protein plan, were twice as successful in maintaining lean muscle mass."

A 2000 study performed by scientists at the University of Sao Paulo in Brazil indicates that BCAA supplementation could boost immunity and decrease the incidence of illness in athletes during intense exercise sessions. The research showed that triathletes who were supplemented with BCAAs were able to prevent a decline in immunity during intense exercise by maintaining critical plasma glutamine levels before and after the study. The athletes taking the placebo showed a reduction of 22.8% in plasma glutamine levels after a competition. Glutamine levels are a key indicator of immune health, and BCAA intake is a proven way to increase plasma glutamine and thereby enhance overall immune function.[90]

High Alpha-lactalbumin Whey Isolate

When it comes to supplying the body with nature's highest levels of essential amino acids – especially BCAAs – nothing beats a high alpha-lactalbumin whey isolate. Not only are properly manufactured (using low heat and spray drying to ensure high bioavailability) high alpha-lactalbumin whey isolates abundant in naturally occurring bioavailable BCAAs, they also provide the right precursors for manufacturing the most powerful antioxidant within the body – glutathione (GSH).[91]

Aside from its role as a key antioxidant in the body, glutathione has also been shown to modulate immunity, detoxify the body and provide anti-cancer protection. In fact, critical illness is directly associated with a depletion of cellular glutathione levels.[92,93,94]

How about Glutathione?

Properly processed whey protein is also known for its ability to enhance lean body mass and thereby metabolism.[95] But new research shows it can also aid in muscular performance. Research presented in the 1999 *Journal of Applied Physiology* demonstrated that enhanced glutathione levels increased muscular performance and reduced excess oxidative stress in 20 healthy subjects. What makes this study so special is that it is the first to demonstrate a direct improvement in athletic performance by enhancing glutathione levels through a specialized whey protein (by a whopping 35%.)[96]

In order for whey protein formulas to boost glutathione levels at the cellular level, they must be properly formulated using low temperatures. These properly formulated whey proteins contain very high levels of the amino acid cysteine, occurring in the form of cystine (a dipeptide). It is only through the delicate (heat sensitive) structure of these dipeptides (cys-cys and glu-cys) that glutathione levels can effectively be increased within the cell.[97] During the pasteurization or processing of the whey protein, temperature differences of only a few degrees is enough to denature or break apart this delicate structure, negatively affecting its glutathione raising abilities.

More benefits

The bioavailability and bioactivity of the various fractions of protein found within your whey supplement become the separating factors between an exceptional quality low-temperature (cold-processed) microfiltered whey isolate and a cheap, inferior one. Aside from this fact, the alpha-lactalbumin portion of your whey supplement is also of paramount importance. Alpha-lactalbumin is one of the most (if not the most) important components of mother's breast milk.

Studies on alpha-lactalbumin indicate that it is without question one of the most effective forms of nutrition required for overall growth and development (one of the main reasons it is found in such high quantities within breast milk). Alpha-lactalbumin also contains nature's highest levels of tryptophan, an essential and often lacking amino acid needed for maintaining levels of the brain chemical serotonin. Serotonin is essential for effective appetite control, sleep regulation and mood improvement – especially under stress.

In fact, serotonin levels have been shown to decline precipitously under stressful conditions, which is one of the reasons we experience excessive cravings, sleep disturbances, depression and memory loss during stressful events. Ground breaking research presented in the *American Journal of Clinical Nutrition* in 2002 showed that only whey proteins containing high levels of alpha-lactalbumin were able to improve brain performance in stress-vulnerable subjects by increasing brain tryptophan and serotonin levels.[98]

Breakthrough research presented in the prestigious *American Journal of Clinical Nutrition* in May 2005 shows that by consuming an evening milkshake containing a special alpha lactalbumin – found in Fat Wars Ultimate Protein – whey isolate product, healthy people could get a better night's sleep and awaken feeling "alive," refreshed and energetic the next morning. After giving 28 healthy young adults with mild sleep problems alpha lactalbumin protein before sleeping, Dutch researchers found that the protein caused a 130 percent increase in tryptophan levels, which seemed to be responsible for a deeper and more restorative sleep. The researchers were also amazed that the participants who had mild sleep problems showed a marked improvement in morning alertness and experienced significantly reduced sleepiness the following day.[99]

Aside from low serotonin levels, one of the major reasons people have such a hard time sticking to their dietary programs is because of enhanced hunger signals. Research indicates that whey proteins containing high levels of specialized proteins called glycomacropeptides, or GMPs, are able to help people feel less hungry between meals by stimulating a hormone called cholecystokinin, or CCK, responsible for controlling our hunger responses.[100,101]

If your goal is to increase your overall metabolism, make your fat loss goals a reality, and enhance your overall immunity and energy levels, research shows that a properly processed high alpha-lactalbumin whey isolate is a must. (Please see Appendix I for the highest alpha whey isolate available today.)

Essential Fatty Acids

Did you know that if you continue to consume the wrong types of dietary fat, you could very well be setting yourself up for continual fat storage and a state of perpetual lethargy? The good news is, if you consume a proper balance of the right type of dietary fats, you could very well experience continual fat burning as well as increased energy levels. Your body literally thrives on beneficial fats, or as nutritional scientists like to call them, essential fatty acids. The key to optimal health and vitality is to supply the right amount of these essential fats to every one of your cells.

Essential fats keep your cells happy and encourage them to burn more body fat, but first they must transform into specialized messengers (called prostaglandins) that increase hormonal communication between cells. These essential fats are referred to as omega-6 and omega-3 fats.

Omega-6 fats, also referred to as linoleic acid, are found primarily in polyunsaturated oils like safflower, sunflower, cottonseed, corn, walnuts, wheat germ and soybeans. Your body requires between three and six percent of your total calories per day (about 1 tablespoon) of these omega-6 fats.

Omega-3 fats, also referred to as alpha-linolenic acid, are found in large amounts within flaxseed, Perilla oil, and in lesser amounts in canola and hemp seed. The main benefits from omega-3's come from their ability to convert into two health-promoting fats – eicosapentaenoic acid (EPA) and docosahexaenoic acid (DHA) – which are found preformed in cold water fish.

With the exception of certain types of fish, the majority of our industrialized diets are pretty much void of the omega 3's, yet abundant in the omega-6. In fact, over 2,000 studies have demonstrated myriad disorders associated with this omega-3 deficiency. The problem is so vast that many researchers believe that well over half the North American population is deficient in omega-3 fatty acids.

The integrity of your cellular metabolism is also highly dependant upon these special fats. In fact, the omega-3 fatty acid EPA has been shown to lower blood levels of fats by increasing the transport of fatty acids into the cells for fat burning (thermogenesis). Omega-3s are so important to metabolic function that many researchers believe our current obesity epidemic is at least in part due to this deficiency.

Fibre – the Missing Metabolic Key

High-fibre intake is linked to a lowered incidence of intestinal disorders, cancer, heart disease, gallstones, diabetes and obesity.[102] Fibre is found in the indigestible portion of plant foods: the skin, peel, stalk, seeds, hull or germ. Its main roles are to absorb moisture in the body, add bulk to the feces and act as fuel for beneficial bacteria.

Even though fibre is derived from carbohydrate sources, the body lacks the necessary digestive enzymes to break it apart and extract its calories, thus making it a true calorie-free food, one which helps maintain metabolic activity. There are two classes of fibre. Soluble fibre (pectins, gums and mucilages) – the sponge – is found in oats, barley, peas, legumes, certain fruits, and psyllium (pronounced sil-e-um). Soluble fibre forms a gelatinous mixture with liquid in the digestive system. In doing so, it can absorb some of the body's excess cholesterol. What gives soluble fibre its grand appeal is its ability to slow the release of glucose into the bloodstream following a meal, thereby lowering our overall insulin response and fat-storage mechanisms.

Insoluble fibre (cellulose, hemicellulose and lignins) – the mop – is found in leafy vegetables, root vegetables, and whole grains like oats and barley. This type of fibre passes through our digestive tract largely intact. Insoluble fibre is responsible for supplying the bulk, or roughage, that keeps foods moving through the digestive system unobstructed, helping to keep us "regular".

I receive emails daily from Fat Warriors who question what percentage of each class of fibre they should consume daily. My answer is always the same: the most important thing to remember when it comes to consuming fibre is the more you consume of any and all types of fibre, the better off your metabolism will be for it. You should always aim to consume between 30 to 50 grams of fibre each and every day. Yet, according to research presented in *Nutrition Review* in 2001, the average dietary fibre intake in the United States is approximately 15 grams per day, but it is probably much lower than this.[103] Without proper amounts of fibre in your diet, the food you consume cannot pass quickly enough through your intestines and your colon, which creates a backlog of chemical waste, which can eventually poison you.

Dietary fibre acts in numerous ways to help maintain metabolism and lower the incidence of obesity. One of its most powerful effects lies in its ability to help lower excess insulin by displacing some of the dietary sugars (from carbohydrates) that are – in a fibre-depleted state – quickly absorbed in the small intestine.[104] Insulin is the chief regulating hormone of fat storage. High resting levels create an enhanced environment for continual fat storage, all the while preventing fat release.

Research suggests that by slowing the release of sugars into the bloodstream – thereby preventing a high surge in insulin – a high-fibre diet can also decrease circulating free fatty acids, LDL cholesterol and triglycerides, while lowering a person's overall risk of cardiovascular disease and obesity.[ciii] Gel-forming fibres such as those found in guar gum, are particularly effective in reducing the elevated negative LDL cholesterol without changing the positive HDL cholesterol levels.[105]

Research presented in the *International Journal of Obesity* indicates that an increased intake of dietary fibre is useful for the treatment of both obesity and diabetes mellitus.[107] Fibre interferes with the body's ability to store fat by upgrading insulin receptor sensitivity, especially within the muscle cells. When insulin receptors in muscle cells are activated, there is correspondingly low insulin receptor activity in the fat cells, and more sugars are diverted to muscle cells as opposed to fat cells. Studies have shown that adding 5 to 30 grams of fibre (especially gel-forming ones like guar gum) per day to a healthy diet leads to an increased satiation (sense of fullness) response, which ultimately reduces the desire to consume extra calories that contribute to body fat stores.[108]

After reviewing several weight loss studies, researchers from Tufts University concluded that the addition of only 14 extra grams of fibre per day should lead to a 10 percent reduction in a person's overall calorie intake.[109] This simple strategy can easily create a four-pound deficit in body fat in as little as four months. Any way you roll the dice, your overall health profile can only be enhanced with the addition of healthy food based fibres, especially those coming from fruits (primarily berries) and vegetables (the non-starchy variety).[110] If you are not consuming sufficient quantities of fibre each day, perhaps it's time you started supplementing your daily intake. (Please see Appendix I for a revolutionary organic fibre complex.)

Healthy bowel recommendations

- Try to obtain the majority of your daily fibre intake from your diet: organic vegetables and fruits provide the best sources, followed by bran, oatmeal, seeds and nuts.
- Drink a minimum of eight, 8 oz. glasses of filtered water every day, and more if you exercise.
- Perform proper exercise to help peristalsis (the natural smooth muscle function of the intestines.)
- If you feel the need to eliminate, don't wait!
- Supplement with high-quality (preferably organic) fibre formulas that contain natural fruit and vegetable fibres.

Carnosine

Carnosine is a dipeptide, a mini-protein comprised of two amino acids (beta-alanine and histidine), produced naturally by numerous cells in the body including muscle, heart and brain cells.

Besides its role in keeping muscle cells contracting, preventing muscular fatigue (by removing excess lactic acid), modulating brain function (by sensitizing neurons to various messages and protecting them from over stimulation) and regulating heart function (by regulating calcium ions), it is one of the most exciting anti-aging nutrient discoveries ever.

Data from studies carried out in 1994 and in 1999 show that carnosine has the ability to cause cells to revert to their once youthful state.[111,112] After adding carnosine to the cell medium, researchers from the Australian Commonwealth Scientific and Industrial Research Organization (CSRIO) were not only able to drastically extend the life of both lung and skin cells, but also were able to keep them in a youthful state right up until senescence (cell death). Most amazing was that when these cells were moved back into a regular cell medium, they instantly started to show signs of aging.

When it comes to stopping the destructive effects of age – especially in skin, brain and eye aging – nothing seems to beat the potential of carnosine. In numerous studies, carnosine was observed to be a powerful agent in stopping both cross-linking and the formation of Advanced Glycosylation End-products (AGE) in various tissues.[113,114] The cross-linking of proteins not only causes the breakdown of your skin, but it also plays a powerful role in inflammation while contributing extraordinary amounts of free radicals in the body. In fact, AGE proteins often produce up to 50 times more free radicals than normal body proteins that have not been damaged by sugars (non-AGE proteins).

One of the most destructive pro-aging and protein-destroying free radicals is the hydroxyl, a radical so powerful, it can destroy DNA. One study proved that carnosine has the ability to significantly protect DNA from oxidative damage.[115] In fact, it was the only substance able to do so! Considering carnosine levels in our bodies decline by 63% between the ages of 10 and 70, consider supplementing this invaluable substance. The recommended dose is 50-150 mg per day.

CHAPTER 9

EXERCISING YOUR METABOLISM

We were all given an incredible vehicle at birth, a miraculous machine capable of moving us around at peak proficiency. In youth, the human body is amazing indeed. Given the proper training and fuel, it can be programmed to achieve remarkable feats of speed, agility, flexibility and strength. But unfortunately, many of us never give a second thought to the inevitable downward spiral that a sedentary lifestyle eventually brings.

You'll remember that the fuels you consume create energy within tiny little power plants called mitochondria, which are present in almost every one of your trillions of cells. They are responsible for manufacturing a high-energy compound called ATP. Without ATP, you couldn't blink your eyes, let alone have the energy to read this page. The health of your body and the efficiency of your cellular metabolism is completely dependant upon the numbers and the activity of these little power plants within the cells. In fact, if you were to look at the muscle cells of a typical, sedentary North American couch potato, you would notice a number of bean-shaped structures. These are the mitochondria. Now, if you were to look at a muscle cell of a healthy, active person of the same age, you would once again see mitochondria, but you would see a whole lot more of them. It is not uncommon for the muscle cells of an active person to contain up to ten times the number of these energy producing power plants.[116] Why is this important to you? Imagine what you could accomplish if you had ten times the energy that you presently have!

Get Moving

By now, you must realize the vast importance in retaining lean body mass (muscle tissue and strength) if your goal is to achieve an efficient metabolism, one that will allow you to live a long, lean and healthy life. Among other things, proper exercise is proven to retard aging[117]; reverse obesity[118]; slow or reverse Type II diabetes[119]; reduce cholesterol[120], high blood pressure[121], stress levels[122], insulin levels[123] and the risk of osteoporosis[124]; keep us from losing vital hormones[125]; increase utilization of oxygen[126] and boost self-esteem.[127]

If there ever was a metabolic "magic pill," it would come in the form of regular proper exercise. But, what is proper exercise? It's the perfect balance of progressive resistance activity, synergized with short duration – but high intensity – cardio activity. Resistance training triggers the body to repair and rebuild itself (anabolism). The irony is that resistance exercise achieves this result (enhanced repair) by first causing your body to break down (catabolism). It is during rest from the resistance portion of exercise that your body switches into a highly anabolic environment. The more you enhance your muscle function, the better your metabolism and the more fat you can burn 24 hours a day – even while you sleep. So, exercise is essential to add muscle to your frame or retrain the muscle you currently carry to become more metabolically active.

The Right Way to Exercise your Metabolism

Most people, especially women, try to lose excess body fat through cardiovascular exercise alone. Unfortunately, research dictates that this is the wrong approach.[128,129] If you are going to commit yourself solely to aerobic training, then you will be carrying around that spare tire for a lot longer than you'd like. Studies show that too much cardio performed in the absence of resistance training can be detrimental to your metabolic engine – your muscle tissue. Research has confirmed that high-intensity cardio activity cannot maintain muscle mass on its own. Don't misunderstand me; cardio exercise is not the enemy. In fact, in order for you to achieve your ultimate fitness goals, it is imperative that it is combined with weight resistance training. Be careful, however, because excess cardio activity can actually eat away at your muscle tissue[130], and that can spell disaster not only to your fat loss efforts but also to your metabolism.

According to researchers at McMaster University in Hamilton, Ontario, the magic of effective exercise lies within the balance, duration and intensity of both anaerobic and aerobic activity.

In a landmark study published in the *American Journal of Cardiology*, aerobic training was compared to a combination of aerobic and resistance (weight) training. Two groups completed a 10-week exercise program of 75 minutes twice weekly. One group completed 75 minutes of just aerobic exercise, while the other completed 40 minutes of aerobics plus 35 minutes of weight training. Training time was identical. At the end of the study, the aerobics only group showed an 11% increase in endurance but no increase in their strength. The group that completed the combination of aerobics plus weight training showed a massive 109% increase in their endurance and a 21-43% increase in their overall strength.[13] There are many other studies that prove resistance training, combined with low-impact cardio, is superior to either one alone.

The bottom line is this: in order to stop our bodies from declining, we first have to engage in a progressive resistance program. Weight training is the only scientific way to increase the lean muscle and strength that you lose through aging. Having said this, it's important to balance both anaerobic (weight training), and aerobic (walking, etc.) exercise in order to affect as many hormonal pathways as possible. The two together become your "magic pill" for lifelong metabolic proficiency.

CONDITIONS OF DYSFUNCTIONAL METABOLISM

By now you should have a pretty good understanding about the importance of maintaining a functional cellular metabolism. You should also be well aware of the tools needed to awaken and maintain an optimal metabolism. Since I have done all that I can in providing you with the information necessary for metabolic success, the rest is in your hands. After all, what good is a toolbox if it is never opened?

In this section you will find an explanation of the following 23 conditions (listed in alphabetical order):

1. Adrenal Health
2. Allergies
3. Arthritis
4. Bone Health: *Osteopenia and Osteoporosis*
5. Brain Health: *Alzheimer's, Parkinson's And Dementia*
6. Cancer Prevention
7. Chronic Pain
8. Diabetes
9. Fibromyalgia/Chronic Fatigue Syndrome
10. Gut Health: *Acid Reflux/Heartburn, Candidiasis. IBS, Crohn's disease, Colitis and Celiac*
11. Heart Health
12. Her Health: *Menopause*
13. His Health: *Andropause*
14. Libido
15. Liver Health
16. Lung Health: *Asthma and Bronchitis*
17. Oral Health: Gum Disorders
18. Metabolic Syndrome
19. Obesity: *The True Epidemic*
20. Prostate Health

21 Skin Health: *Acne, Eczema And Psoriasis*
22 Sleep
23 Thyroid health: *Hypothyroidism*

At the end of each one of these conditions, you will find a recommended **Nutrient Chart** with listed recommendations of individual nutrients and pre-existing formulations (comprising many of the recommended nutrients in each section) along with dosage considerations. The Nutrient Charts are designed this way to make it easy for you to look for synergistic combinations of effective nutrients for each specified condition.

These nutrient recommendations are not – *in any way* – to be taken as a prescription for any condition, as only your current health care professional can provide this information. Rather, they are provided as a guide to the latest nutrient protocols from various scientific organizations around the world. It is important to note that the suggested dosages are not to be mistaken for *therapeutic dosages* in any way, as therapeutic dosages (**prescribed by a natural health care professional only**) are usually quite a bit higher than the ones listed. The suggested **Nutrient Chart** dosages are instead recommendations for prevention, NOT DISEASE!

Over the years, I have often found myself providing information to confused nutrient consumers on the best ways to take their various nutrient combinations. People are often not told what foods – *if any* – would be best to combine their nutrients with. Some nutrients work best with fat soluble foods, some away from dairy foods, some on an empty stomach, some *with* or *without* food, some first thing in the morning and some last thing before bed.

Below you will find my **Nutrient Optimization Chart**, which is designed to take all the confusion surrounding these topics away from you. All of the recommended nutrients and formulations described in the "various condition related" Nutrient Charts below appear in the **Nutrient Optimization Chart** to make it as simple as possible for you – enjoy!

Nutrient Optimization Chart

	With Food	Without Food	With/or Without Food	Awa from Dairy	With fat contain-ing Food	Day or evening	A.M.	P.M.
Ultimate Protein Energy Shake				X				
FibreLean			X	X				
Ultimate Longevity (with carnosine)		X					X	
Ultimate Her Energy			X			X		
Ultimate Male Energy			X			X		
Ultimate Lean Energy		X					X	
Ultimate Anti-Stress			X					X
SierraSil (LaniFlax)		X		X		X		
Immuno-Care		X		X		X		
X2O			X			X		
Multi-vitamin/mineral complex					X		X	
B vitamins (high-potency)	X						X	
Vitamin C (as mixed ascorbates)			X			X		
Vitamin E (as mixed tocopherols)					X	X		
Co-Q10					X		X	
Alpha lipoic acid (ALA) R+ form			X				X	
L-carnitine (and acetyl-L-carnitine)		X					X	
L-glutamine		X				X		
Calcium	X					X		
Magnesium	X					X		
Vitamin D3 (Cholecalciferol)					X	X		
Zinc (as picolinate or citrate)		X						X
Selenium (as LeafBrand or Se-Methylselenocysteine)			X			X		
Chromium (as LeafBrand or polynicitonate)			X			X		
Probiotics		X				X		
Saw palmetto		X				X		
Ziziphus spinosa seed extract		X						X
NADH (enteric coated)		X					X	
Melatonin		X						X

A Word about Green Drinks

I have often believed that nature provides the best nutrient "bang for your buck." By this I mean that we can never do better than what already occurs in nature, at least where the human body is concerned. Having said this, I also realize that according to the latest statistics, we consume but a mere fraction of the total daily recommended 6-10 servings of fruits and vegetables. Aside from this fact is the reality that whatever fruits and vegetables we do consume, if not organically produced, are quite deficient in the nutrients the human body thrives upon. In fact, according to a report presented in the *Journal of the American College of Nutrition* in December 2004, levels of six nutrients – protein, calcium, iron, potassium, riboflavin and vitamin C – in non-organically produced vegetables and fruits, have declined dramatically since 1950.[1]

In the recommended Nutrient Charts below, I do not include green drinks. However, I do often recommend supplementing your diet with properly formulated 100% *certified organic freeze-dried* green food concentrates, especially if your diet is currently low in *organic* produce. But before you go out and purchase just any green food concentrate, please be aware that some provide much more nutritive value than others (please see Appendix I for the one I personally use).

Here's what to look for: Green food concentrates should provide 100% *certified* (USDA) organic freeze-dried ingredients, coming primarily from fruits and vegetables including: Spinach, Blueberry, Kale, Parsley, Cranberry, Red Cabbage, Green Cabbage, Broccoli, Brussels Sprouts, Okra, Papaya, Rose Hips, Pomegranate, Grapes, Oats, Brown Rice and spices like, Turmeric, Ginger, Cinnamon and Chicory, cultured in a base of probiotics.

23 A-Z CONDITIONS OF DYSFUNCTIONAL METABOLISM

one

ADRENAL HEALTH

Continuous stress has been linked to North America's five leading causes of death: heart disease, cancer, lung disease, accidents and cirrhosis of the liver. Author and stress researcher Dr. Kenneth Pelletier has contended that, in America, between 80 and 90% of all illnesses can be linked to stress and that 75 to 90% of all visits to the doctor are for stress and anxiety-related concerns.[2]

In North America, many people don't even know we have adrenal glands, let alone know how important they are to our health. Yet, anyone who experiences any sort of stress has intimate knowledge of the effect of our adrenal glands. (Thank goodness there is no stress in your life!)

No one has to tell your body what to do when it perceives fear. It is an instant response, hardwired from primitive times in an area of the brain called the amygdala. The amygdala, an almond-shaped region in the forebrain, controls and directs the alarm mechanism (*fight or flight*) in both the central and autonomic nervous systems.

The amazing thing is that any reaction to the fear (such as freezing) takes place within a few thousandths of a second.[3]

Anatomy Lesson

We have two bean-shaped adrenal glands. Each one sits atop a kidney. The glands are divided into two parts: the cortex and the medulla. The medulla is actually an extension of the central nervous system. In response to stress, whether it is actual or perceived, the medulla secretes two powerful hormones – norepinephrine and epinepherine (adrenaline) – that stimulate the "flight or fight" reaction. During a stressful event, cardiac output is affected, along with blood flow to the muscles, lungs and brain. You might notice that your breathing becomes shallow, and you experience tightness in your stomach or a dry throat.

The adrenal cortex, on the other hand, produces corticosteroid hormones including glucocorticoids and mineralcorticoids that are a factor in the regulation of inflammation, blood electrolyte levels, behaviour, reproduction, excretory and immune system functions, as well as regulating metabolism. The major glucocorticoid is cortisol, which is important in dealing with chronic stress. Chronic stress isn't the short-term reaction you experience when a child jumps out from behind the couch, squealing "boo!" Instead, chronic stress is the lingering anxiety that involves balancing the chequebook, worrying about the health of a loved one, or holding on to your job. In order to protect the brain (which can only function on a steady supply of glucose), cortisol causes the liver to free up glucose, while also limiting the ability of our tissues to take up glucose. It does this by stimulating the catabolism of protein from all tissues, except the liver. So, high stress leads to high cortisol, which ultimately leads to the breakdown of lean muscle, and since lean muscle to a large extent controls the rate at which your body burns calories over a 24 hour period, cortisol has the potential to destroy your metabolism.[4] Interesting.

The cortex also produces a small amount of testosterone and other male sex steroids including dihydrotestosterone (DHT). Known as androgens, these hormones are involved in gene regulation, behaviour, sex differentiation and anabolic metabolism. The adrenal glands also produce the hormone dehydroepiandrosterone (DHEA), also known as the "mother" hormone because it is a precursor for the manufacture of many other hormones including cortisone, estrogen, progesterone and testosterone. DHEA works as a "buffer" hormone, and is particularly beneficial to women experiencing the hormone swings of peri-menopause.[5]

Now that you have an understanding of the role of your adrenals, you can see the importance of keeping them strong and healthy. But we live in a crazy, fast-paced world, where high levels of anxiety have become the norm. And we are working our stress glands to the point of exhaustion. Continual production of the stress hormone cortisol can lead to compromised immune function, meaning that you are more susceptible to the latest, most fashionable virus or flu that is circulating around town. You might also find that it takes you a lot longer to recover from an illness than it used to. Elevated levels of cortisol are also linked to memory loss, infertility, chronic fatigue syndrome and fat accumulation, particularly around your middle.

Stress – The Caffeine Connection

Think you need that cup (or carafe) of coffee to keep you motivated as you go about your day? If so, you likely don't realize that you're contributing to your stress levels and wreaking havoc on your adrenal glands. A study of the effect of caffeine on heart rate, blood pressure and stress hormone levels found that caffeine significantly raised average blood pressure and reduced average heart rate by 2 beats per minute in adults. I don't need to remind you that stress and high blood pressure can contribute to an increased risk of heart disease. In addition, caffeine may exaggerate the impact of stress on the adrenal system.[6] In other words, coffee ages you, especially if your adrenals are already in a compromised state.

Other familiar symptoms of adrenal exhaustion include, (but are not limited to): low energy, heart palpitations, short temper, dizziness upon rising, shortness of breath, hypoglycemia, neck or shoulder tension, cold feet and knee problems. Do you notice how many of these symptoms are similar to the ones we associate with getting older?

Supporting the Adrenals

The best way to give your adrenals a rest is to decrease the stress in your life. Take long walks with a loved one, sign up for a yoga class, or learn how to meditate. Re-arrange your life to be less demanding. Avoid stimulants like sugar and coffee, as well as depressants like alcohol, and eat a diet focused on fresh fruits and vegetables, along with lean protein, nuts and seeds. As you will see, various nutrients, in supplement form, have also been shown to help tired adrenals bounce back, many in record time.

WONDER HOW YOUR ADRENALS ARE DOING?

Try the Orthostatic Blood Pressure Test (orthostasis simply means standing upright.) All you have to do is compare your blood pressure while lying down to your blood pressure when you are standing. When you are healthy, your blood pressure is usually lower when you are lying down because you aren't expending any effort. When you stand up, your heart has to work harder, so typically your blood pressure increases. If you are experiencing adrenal exhaustion, however, your blood pressure drops when you change from a lying position to a standing position. To investigate your adrenal function, take your blood pressure while lying down. Then stand up and immediately measure your blood pressure again. If your blood pressure drops when you stand, suspect adrenal exhaustion.

Adrenal Protection Nutrient Chart

Nutrient	Suggested dosage
Ultimate Protein Energy Shake™	1 serving twice daily
Ultimate Anti-Stress™	2-4 capsules prior to sleep
Multi-vitamin/mineral complex	Follow label instructions
B vitamins (high-potency)	Follow label instructions
Vitamin C (mixed ascorbates)	500-2,000 mg/day
X_2O mineral sachets	1-2 sachets twice daily in water

two

ALLERGIES

The immune system is designed to protect the body from threat. That threat might come in the form of a virus, bacterium, fungus or parasite. A healthy immune system goes about its business seeking out and destroying harmful agents that have found their way into your body without you having to give it a second thought – that is, until a cold or flu drags you crawling back to bed. For some people, however, severe immune system deficiencies leave them open to all sorts of dangers. For others, the immune system is on overdrive, attacking particles that typically cause no harm. Allergies fall into this second category.

To understand what causes an allergic reaction, we need to outline what the immune system is and a few examples of how it works. (As you read through this explanation, remember that every one of your trillions of cells has its own metabolism.) Unlike the digestive system, for example, the immune system is not located in a particular region of your body. Instead, immune cells are located in every area of your body and are designed to perform various defense services. Coursing throughout your body, some of these cells identify a threat, others dismantle the threat, and still others clean up the mess. These threats are collectively known as antigens.

And the immune system is smart. It only has to identify and defeat an antigen once and details of the encounter are encoded into memory cells that are forevermore reproduced. If the memory cells or antibodies meet up with that antigen again, the antigen won't last long. That's why you'll never get the same cold virus twice. (Hard to believe, I know, but every time you get the sniffles, you can thank a different virus.) Your immune system turns up your core temperature to burn antigens, fills your head with mucus to drown them out of your system, and your sore throat is thanks to the additional blood supply filled with immune cells hanging around at the logical port of antigen entry. The misery you might feel when you are under the weather is the result of your effective immune system, and the health of that immunity is dependant upon the immune cell's metabolism!

Too smart for your own good

With allergies, the same effective system is at work... it's just slightly off target. Instead of identifying and attacking an antigen, the immune system incorrectly believes that something rather harmless, like pollen or peanuts, needs to be obliterated. In the case of allergy, the "threat" is called an allergen. The first time the body is exposed to an allergen, the immune cells make a huge batch of antibodies, which attach to cells called mast cells. Mast cells are located in the linings of the nose and gastrointestinal tract, lungs, skin and tongue. When the allergen is encountered again, the antibodies prompt the mast cell to release histamine and other inflammatory chemicals that cause the wheezing, sneezing, tears and runny nose considered common symptoms of allergy. But allergies can make themselves known in other ways, as well. Other symptoms include eczema and contact dermatitis, itching, hives, swelling, lower blood pressure, difficulty breathing, asthma attacks and in serious cases death. (See Part Two, Condition 15, Lung Health.)

The Statistics

There are as many potential allergens as there are people. About 75% of allergy sufferers consider environmental allergies to be their primary allergy. (Unfortunately, many people with allergies are allergic to more than one thing.) The most common environmental allergy triggers are pollen, mold spores, dust mites, cockroach allergen and animal dander. Cat dander ranks as the most common pet allergy, affecting approximately 10 million people in the U.S. alone. Food is a common trigger, and occurs in roughly 1 to 2 percent of adults and 3 to 8 percent of children. About 90 percent of all food allergies involve milk (and dairy products), eggs, peanuts and tree nuts, wheat, soybeans, fish and shellfish. Genetics play a role in allergies: if one parent has allergies, the offspring has a 33% chance of being allergic. Odds for children increase to 70% if both parents are allergic. With yearly costs beyond $18 billion dollars, allergies are the sixth leading cause of chronic disease in America and are believed to affect the lives of nearly 50 million people in the U.S.[7]

So what can you do?

Besides avoiding the allergen that gets you into trouble in the first place, make sure that you have addressed any underlying problem with your digestive system, particularly if you experience food allergies or food sensitivities. (Sensitivities differ from allergies in that they seem to be dependent on the quantity and frequency of the exposure – meaning sometimes you can eat them and they don't bother you at all – as well as the severity of the reaction.) Consider whether you have an imbalance in your intestinal flora, particularly with *Candida*. *Candida* is often a factor in food allergies as well as environmental allergies. (See Part Two, Condition 10, Gut Health)

Carolee Bateson-Koch DC, ND, in her book *Allergies: Disease in Disguise*, suggests that we often become addicted to foods that we are allergic to. It's not a coincidence, therefore, that the most commonly eaten foods are also the most common allergens. According to Bateson-Koch, cravings and insatiable hunger are sure signs of withdrawal, and withdrawal is an indication of addiction. Insatiable hunger is often tied to obesity. If you crave particular foods, or can't seem to stop eating them once you get started, consider eliminating those foods from your diet. You might find that you will lose about ten pounds simply by removing these possible food allergens from your menu![8]

In other food-related news, research shows that the fatty acid arachidonic acid found in red meats, organ meats, pork meats and high fat dairy products are associated with hay fever.[9] Arachidonic acid is a precursor to inflammatory chemicals in the body called leukotrienes, which are up to 10,000 times more potent than histamine. You might want to limit your intake of these foods if you have allergies and instead focus on anti-inflammatory gamma linoleic acid (GLA) from borage oil and eicosapentaenoic acid

(EPA) from fatty fish. Fish oil supplements work very well if you are not a fish lover. Alcohol might also go on your watch list: it is involved in several hypersensitivity reactions, including asthma, food allergy and anaphylaxis (a severe reaction that can lead to death.)[10] If you think you can't live without alcohol, reread the previous paragraph referencing addiction!

Whether or not you have food allergies, focus your diet on whole foods rather than prepared foods. You want to limit your exposure to chemicals and possible allergens. Eat a variety of fruits and vegetables, particularly those high in vitamin C. Vitamin C helps to break down histamine, combat inflammation, offer antioxidant protection and protect lung function.[11]

Remember all of the cells involved in your immune system, and how they function in the case of allergies. Nourishing your cells might not make you any less allergic, but you will feel better and recovery time after an episode will be faster when you focus on your metabolism!

Allergy Prevention Nutrient Chart

Nutrient	Suggested dosage
Ultimate Protein Energy Shake™	1 serving twice daily
ImmunoCare™	1 capsule/day between meals
Multi-vitamin/mineral complex	Follow label instructions
Probiotics	Follow instructions
Vitamin C (mixed ascorbates)	500-2,000 mg/day
X_2O mineral sachets	1-2 sachets twice daily in water

three

ARTHRITIS: Joint health

When most think of a disorder associated with aching and often inflamed joints, the first thing that comes to mind is arthritis. However, arthritis is actually a disorder representing over 100 rheumatic diseases (those associated with pain, inflammation and often limited mobility.) The most common rheumatic disease is osteoarthritis. Between 70 and 80% of the population over the age of 50 are affected by this disorder. This is, in large part, due to the natural (or unnatural) wear and tear of joint tissue that develops through the aging process. In fact, if we all lived long enough, the wear and tear of the aging process would eventually have every one of us living with osteoarthritis; that is, if we do not take the necessary steps to maintain the anabolic metabolism through proper nutrition and supplementation, catabolism will eventually get the better of us – especially our joints!

In order to understand how osteoarthritis develops, we must first understand the action of joints. Joints are connection points where two or more separate bones are joined. Your body contains over 140 joints and every one of them is designed to offer maximum protection to the ends of your bones by acting as shock absorbers that reduce friction and allow easy gliding and rolling of joint surfaces against one other. To give you an idea of just how efficient joints are at reducing friction (when they are healthy), researchers from Harvard School of Medicine have calculated the friction of joints to be lower than that of ice gliding on ice.[12]

Joints offer their unique bone protecting characteristics through the aid of a flexible, rubber-like material called cartilage. Cartilage is comprised of three primary substances: water, collagen and proteoglycans (pro-tee-o-glycans). Water is the main substance in cartilage, comprising anywhere from 60 to 85% of the matrix, followed by collagen (the structural protein of cartilage) at 10 to 20% and proteoglycans (large molecules of protein and sugar) at 5 to 10%.[13]

Collagen is the most abundant structural protein in the body and forms the foundation of cartilage by acting as the framework that holds the proteoglycans in place. Collagen also provides the tensile strength to the cartilage as well as elasticity and shock absorption.[14] Proteoglycans are responsible for attracting and binding to water, which is extremely important to the functioning of cartilage. This is what allows cartilage to offer resistance to compression. To give you an idea of just how much compression is exerted upon cartilage at any given moment, every time we take a single step, our cartilage experiences the equivalent forces of a 300-lb person hanging from a ledge by a fingertip.[15] And to think the average person takes about one million steps each year – ouch!

Healthy, cartilage cells (also called chondrocytes) are in a continuous anabolic state of renewal through the construction of new collagen and proteoglycans. Aside from manufacturing new materials for cartilage integrity, chondrocytes also produce specialized enzymes responsible for breaking down and recycling collagen and proteoglycans. Problems begin to mount when metabolic activity of chondrocytes becomes predominantly catabolic. This means that for whatever reason (poor nutrition, sleep deprivation, aging, injuries or regular NSAID use), too many breakdown enzymes are produced and all rebuilding activity comes to an abrupt halt. Once this catabolic condition is in place, cartilage destruction in the form of osteoarthritis becomes the end result.

Inflammation is often at the route of cartilage destruction, and many of the common treatments or anti-inflammatories seem to exacerbate this condition. For instance, researchers from the Mathilda and Terence Kennedy Institute of Rheumatology, London, England, have discovered that the pro-inflammatory cytokine TNF-alpha is often the main culprit when it comes to stimulating and maintaining inflammatory messengers that degrade cartilage. The researchers found that TNF-alpha stimulates the production of cartilage destroying cytokines like IL-1 and IL-6, and when TNF-alpha was effectively neutralized; other proinflammatory cytokines were also inhibited.[16]

In numerous studies, TNF-alpha has also been shown to play a prominent roll in the destruction of cartilage where rheumatoid arthritis (RA) is concerned. Unlike osteoarthritis, which is caused from wear and tear, rheumatoid arthritis is believed to be an autoimmune disease, which causes the body's immune system to produce antibodies against human tissue. The human tissue in this case is cartilage. Patients who suffer from RA often express high levels of TNF-alpha along with the cytokines it stimulates, and studies also indicate that anti-TNF-alpha agents seem to help control or inhibit many mechanisms of RA. Unfortunately, the majority of mainstream treatments for osteoarthritis these days tend to come from doctors recommending NSAIDs to their patients. The reason NSAIDs pose a potential risk in cartilage disorders is because they often end up creating more damage to the cartilage in the long run. For instance, Australian researchers have found that by blocking the COX-2 enzyme – responsible for the proinflammatory PGE2 – through NSAIDs, cartilage destruction can actually increase due to the overproduction of other pro-inflammatory cytokines.[17]

Nearly twenty years ago, researchers from the Indiana University School of Medicine in Indianapolis discovered that many NSAIDs negatively affect the metabolism of proteoglycans by blocking the chondrocytes ability to produce them. Their research concluded that despite the symptomatic improvement that these drugs produce, cartilage degeneration could be accelerated.[18]

Since NSAIDs do not offer any protection against cartilage damage[19] and may actually cause further cartilage breakdown, it makes sense to look for healthier, side effect-free alternatives. The only way to truly help a condition like osteoarthritis is simply to slow down or halt the destruction to the cartilage. One of the best ways to do this is to provide healing raw materials for anabolic metabolism to take place. An important nutrient to rebuild healthy cartilage tissue is glucosamine sulfate or GLS. Armed with the right food choices, proper supplementation, rest and avoidance of regular NSAID use, the rebuilding of health cartilage is not only possible, it's highly likely! Yes, this means you can beat the pain once and for all. In fact, studies have shown that through the ingestion of effective doses of glucosamine sulfate, anabolic metabolism in chondrocytes can be increased along with the production of more collagen and proteoglycans.[20]

One of the most exciting anti-inflammatory ingredients to come along in a long time happens to be a complete product of nature. SierraSil® is a distinct, naturally-occurring mineral composite containing over 65 naturally occurring macro and trace minerals – which has been documented through clinical research to have powerful anti-inflammatory properties. SierraSil,® which is found only in the high Sierra Mountains, is comprised of a complex of minerals with unique properties unlike any other composite known. In a study published in the October, 2004 issue of the *Journal of the American Nutraceutical Association*, researchers from Case Western Reserve University School of Medicine, were able to significantly reduce (by 68 to 73%) the breakdown of cartilage caused by the pro-inflammatory cytokine Interleukin 1 after treatment with SierraSil®.[21] Researchers concluded that SierraSil® is a viable treatment for joint pain and that its anti-inflammatory properties are likely due to suppression of gene expression.

And move your body

Although even the idea of movement might be painful if you have been diagnosed with arthritis, evidence indicates that exercise will make you feel better. A 2005 study shows that exercise therapy is effective for patients with knee osteoarthritis, hip osteoarthritis and rheumatoid arthritis.[22] Exercise improved quality of life in several conditions including osteoarthritis of the knee, and generally led to improved physical performance. Researchers also concluded that exercise could help to prevent secondary diseases.[23] And, finally, we all know that exercise is an important factor in weight loss. If you are one of those people who are motivated to action by the avoidance of pain, consider this if you have arthritis in the knee and are overweight: a 2005 study published in *Arthritis and Rheumatism* journal found that for every one pound of weight lost there was a corresponding 4-fold reduction in the load exerted on the knee per step during daily activities.[24] Since you log in thousands of steps per day, think how much you can protect your knee joints by dropping a few pounds!

Joint Protection Nutrient Chart

Nutrient	Suggested dosage
Ultimate Protein Energy Shake™	1 serving twice daily
LaniFlex (with SierraSil®)™	Follow label instructions
Ultimate Longevity (with carnosine)™	1-2 capsules/day
Multi-vitamin/mineral complex	Follow label instructions
Glucosamine Sulfate	1,000 mg, 2-3 times/day
Vitamin C (as mixed ascorbates)	500-2,000 mg/day
Vitamin E (as mixed tocopherols)	800 IU daily
X_2O mineral sachets	1-2 sachets twice daily in water

four

BONE HEALTH: Osteopenia and Osteoporosis

When people think of energy and metabolism, the last thing they connect that to is their skeleton. In fact, most people think of their bones as a lifeless structure. However, bones are anything but dead; they are teaming with life and yes, even the cells of your bones have their own metabolism. Bones are highly anabolism and, when healthy, are constantly undergoing renovation. Bone cells (called osteoclasts) do the catabolic work, dismantling and removing old bone. Osteoblasts are in charge of the anabolic bone work, and are constantly rebuilding with new mineral deposits. The function of other cells (called osteocytes) is just starting to be understood, but researchers believe these cells work as sensors in the bone-building process. Compared to the bone-building cells (the anabolic osteoblasts), we have only a fraction of catabolic osteoclast cells. We also have at least ten times as many osteocytes as osteoblasts in our bones.[25]

We continue to build bone density until our late twenties or early thirties, at which point we have reached our peak density. And then our goal is simply to hold on to what we have. When old bone is removed faster than new bone is rebuilt, bones become weaker or "porous," leading to osteoporosis – literally, "porous bone." As a result, bones become brittle and break or fracture easily. Common fracture sites include the wrist, spine and hip. More than 90% of hip fractures are associated with osteoporosis, leading to long-term hospitalization and triggering complications including death. Older adults who have suffered a hip fracture are 5 to 20% more likely to die in the first year following the injury than others in this age group. Osteoporosis affects nearly 30 million North Americans, and although men can also suffer from weakened bones, 80% of those with osteoporosis are women.

Like all of the conditions covered in this book, osteoporosis is not instantaneous: you don't simply wake up one morning with porous bones. There are indications along the way that your bone metabolism is leaning too heavily to the side of catabolism. The first step on the road to porous bones is the thinning of bones, otherwise known as osteopenia (from the Greek "osteo" meaning bone and "penia" meaning poverty.) The World Health Organization classifies osteopenia as a bone density measurement that is one standard deviation below the bone density of a normal adult bone as measured with dual-energy x-ray absorptiometry (DEXA) or the bone density test. Osteoporosis occurs at 2.5 standard deviations below that level.[26]

Major risk factors for bone loss include family history of the disease, thyroid disease, menopause, sedentary lifestyle, and high intake of red meat. Excessive alcohol intake and cigarette smoking are associated risks, as is the consumption of caffeine, particularly in the form of soda pop. Low intake of calcium, phosphorus, magnesium and vitamin D due to poor diet are other avoidable risks. People with a small, thin frame are also at increased risk.

Beyond calcium

Many people mistakenly believe that osteoporosis occurs simply due to a deficiency in calcium. If that were true, osteoporosis could be prevented and cured simply by popping calcium supplements. In reality, many factors play a part in bone building and subsequent bone loss.

In the bone building process, our bodies build a frame for us out of collagen (the body's most abundant structural protein), which then fills in with calcium, phosphorus, magnesium, manganese, boron, strontium, silica, zinc and copper. Our bodies also require adequate folic acid, vitamin B6, vitamin K and vitamin D. Deficiencies in these nutrients can occur either because they are not consumed in sufficient quantities or because there is difficulty in absorption. Many people, particularly the elderly, have low stomach acid and this leads to digestion problems. (See Part Two, Condition 10, Gut Health.) This is especially important to understand as many calcium supplements are made of calcium carbonate, a very difficult form to digest. Other forms of calcium including calcium citrate, aspartate and orotate are more easily absorbed. Problems with the liver metabolism could lead to insufficient production of bile, a component necessarily for the assimilation of all of the fat soluble vitamins including vitamin A, vitamin E, vitamin K and, of course, vitamin D. (See Part Two, Condition 14, Liver Health.)

Vitamin D is essential for the absorption of calcium into bone, and our bodies are capable of producing this vitamin when we are exposed to sunlight for about 20 minutes per day. You can see how this is a problem during the winter months. And a 2004 comparative study showed that university-aged women whose clothing covers their skin experience vitamin D insufficiency and are at increased risk for osteoporosis in later life.[27] Supplementing vitamin D also helps to reduce the risk of falls for the elderly, and a report published in the *American Journal of Clinical Nutrition* suggests that the current recommendation for vitamin D might be too low. The 2005 report suggests that vitamin D intake increase from the current recommendation of 600 IU per day to 800-1000 IU daily.[28]

And again with the hormones

Several hormones are also involved in regulating bone building, including estrogen. As a result, when estrogen levels fall at menopause, the risk of osteoporosis increases for women. Many women believe that supplemental estrogen or HRT will *preserve* bone. However, it does not help to *rebuild* bone. Remember the cyclical nature of bone building as well: if old bone is not broken down through catabolism, it cannot be replaced with stronger, healthier bone. HRT may not be the panacea for osteoporosis after all.

The thyroid hormone thyroxin activates bone catabolism, so long-term, high levels of the hormone for a prolonged period either through hyperthyroidism or medication for hypothyroidism also results in bone loss. Use of medications can also contribute to osteoporosis: corticosteroid drugs decrease absorption of calcium, and antidepressants have been linked to increased risk of hip fractures. Diuretics, antacids, anticonvulsants, warfarin, lithium and several other drugs can also contribute to bone weakening. Inflammation caused by stresses to the immune system can pull calcium from bone, and a sedentary lifestyle also weakens bones.

Stand tall

There is much you can do to protect your bone health, even after a DEXA scan shows bone loss. Some data suggest that the bones of elderly women might benefit from increased intake of protein.[29] High-alpha whey protein isolates offer a superior and easily digestible source of protein to the body, making them an excellent option for bone health.

A 2005 report recommends a nutritionally-balanced diet with the appropriate caloric intake for your activity level, including sufficient calcium, potassium, magnesium and a low amount of sodium. An adequate intake of alkali-rich foods, mostly found in fruits and vegetables, also promotes a beneficial effect on protein metabolism for the bones. Omega-3 fatty acids help to turn down inflammatory influences, while at the same time reducing osteoclastic activity. Researcher's also recommend reduced use of alcohol and caffeine.[30]

Be sure to consume 1000-1200mg of bioavailable calcium daily. Look for your calcium in green leafy vegetables: broccoli, sesame seeds, almonds and sardines. Aim for at least 800 IU of vitamin D daily, and you will find this fat-soluble vitamin in salmon and tuna (great sources of omega-3s).

"To prevent falls, regular strength and balance training are beneficial. It makes sense – you're not as likely to fall if you can maintain your balance."

And don't think you can walk away from this section without a word or two about exercise. Like muscle, bone needs to be used to become stronger, and our bones require weight-bearing exercise. This includes anything that we do upright that forces us to work against gravity. Walking (preferably with hand-weights) and dancing are excellent options, and although swimming is great for the heart, it does little for your bones. According to research published in 2005 in *Osteoporosis International* journal, low-volume/ high-intensity exercise successfully maintains bone mineral density in the hip and spine.[31]

It's also important to consider the importance of muscle strength and agility, especially when considering hip fractures. More often than not, hip fractures occur as a result of a simple fall from standing height or less. To prevent falls, research published in the *Annals of Internal Medicine* states that regular strength and balance training are beneficial. It makes sense: you're not as likely to fall if you can maintain your balance. Try yoga, Pilates and weight training to build core strength and improve your balance!

Bone Protection Nutrient Chart

Nutrient	Suggested dosage
Ultimate Protein Energy Shake™	1-2 servings/day
Multi-vitamin/mineral complex	Follow label instructions
B vitamins (high-potency)	Follow label instructions
Calcium (Glycinate and/or Citrate/Malate)	1,000-1,200 mg/day
Magnesium (Glycinate and/or Citrate/Malate)	600-1,000 mg/day
Vitamin D3 (Cholecalciferol)	800 IU daily
Vitamin K2	15 mg, 1-3 times/day

five

BRAIN HEALTH: Alzheimer's, Parkinson's and Dementia

While we can agree that aging is the result of a deterioration of our metabolic functions, it's also important to understand how the break -down occurs. The Free Radical Theory of Aging suggests that reactive oxygen species (ROS) or free radicals are a major factor.[32]

ROS are the natural by-product of the metabolism of living: they result from breathing, sunlight and eating and involve the oxidation of oxygen. And, yes, oxidation is a synonym for corrosion. Some of our food and lifestyle choices increase free radical production: cigarette smoking, alcohol production and a diet high in saturated fats are known free radical generators. On the sub-cellular level, free radicals are atoms that have lost an electron in their outer shell. This makes the atom "crazy," and it furiously runs around looking for another electron to steal so that it can become stable again. As a result, it creates more free radicals and the chain continues. Along the way, cells become damaged by the constant theft of electrons. Lucky for us, free radical scavengers or antioxidants can put a stop to the rampage of the free radical by donating an electron.

You want to keep this in mind as you consider the health of your brain. Along with being a virtual glucose hog, your brain also uses enormous amounts of oxygen to fuel its metabolic furnaces. Your body delivers at least half of the circulating glucose and about 20% of its oxygen to the cells in your brain. Obviously, then, there is considerable opportunity for free radical damage to take place in your brain, and since the dry weight of your brain is almost 60% fat, in a very real sense your brain can go rancid! The fact remains that the older you get, the lower your natural protection against progressive free radical damage,[33] so the more susceptible you become to age related mental disorders in the form of memory loss, dementia, Alzheimer's disease (AD) and Parkinson's disease.[34]

The Numbers

According to the Alzheimer Association, an estimated 4.5 million Americans currently live with AD[35], a number which has doubled since 1980.[36] In Canada, more than 420,000 Canadians over the age of 65 have AD and other related dementias.[37] The Alzheimer's Association and National Institute on Aging estimate that costs of caring for people suffering from AD are at least $100 billion annually.[35] Parkinson disease (PD) affects 1 in 100 people over the age of 60. Although the average age of onset is 60 years, 5 to 10% of patients with PD are younger than 40.[38]

Alzheimer's (AD)

The most common form of dementia, AD is a progressive brain disorder in which parts of the brain degenerate. Areas of the brain that control memory and cognitive skills are affected first, but cells die in other regions of the brain as the disease progresses. Eventually, those with Alzheimer's lose the ability to communicate, make judgments, and carry out daily activities. In the absence of other serious illness, loss of brain function eventually is fatal.

Parkinson Disease (PD)

Parkinson disease is a progressive disease resulting from impairment and death of cells in the midbrain. These cells are responsible for producing dopamine, which is part of a chain of events that is responsible for coordinating movement. Symptoms of PD become evident when 80% of the midbrain cells have died, and include tremors progressing to bradykinesia (slowness of movement); stooped posture and rigidity, and problems with balance. Many people with PD eventually develop dementia.[39]

Prognosis for Brain Health

Both Alzheimer's disease and Parkinson disease are disorders of the brain's cellular metabolism, as they both involve mitochondrial dysfunction. You'll recall from Chapter 2 that the mitochondria are tiny little energy factories within the cells. The inability of the mitochondria to produce adequate energy coupled with an abundance of free radicals may be responsible for the cell death associated with these and other forms of dementia.[40]

As upsetting as these diseases of aging can be, it is possible to protect the health of your brain, and, not surprisingly, you do this by improving metabolism in your brain cells. In the first part of this book, I covered the importance of feeding your metabolism with protein, carbohydrates and essential fats. Fats are particularly important to your brain, because – as I mentioned above – your brain is made of fat! (See Chapter Six, Dietary Fats) You simply have to be sure that you are providing the right types of fats. Saturated and hydrogenated fats increase cholesterol levels, which can clog arteries and blood vessels that supply nutrients to the brain. (See Part Two, Condition 11, Heart Health) Instead, to protect the health of your brain, eat low fat sources of protein like chicken, turkey, and fish. Boost your brain power with omega-3 fats from cold water fatty fish like salmon and tuna, or use properly formulated fish oil supplements.

Always remember that your brain is made of fats that are subject to oxidation. Therefore, you should also be sure to get adequate amounts of fat-soluble antioxidant vitamins, particularly vitamin E. In fact, low concentrations of vitamin E are common in patients with AD. Several studies show the brain benefits of adding vitamin E supplementation to your daily health regimen.[41] Research conducted at Johns Hopkins University determined that vitamin E and vitamin C in combination is associated with reduced incidence of AD. Results were the same when vitamin C was included in a multi-vitamin supplement along with additional vitamin E.[42] A study reported in the *Journal of the American Medical Association* also found that high dietary intake of vitamin C and vitamin E may lower the risk of AD.[43] Please keep in mind that your body works best on natural forms of vitamin E coming from mixed tocopherols (especially gamma tocopherols.) Make sure you are eating between 5 and 10 servings of fresh fruits and vegetables daily to provide a good supply of antioxidant vitamins, and use supplemental vitamins as required.

Other nutrients that promote brain health include carnosine and co-enzyme Q10. In 2004, the *Journal of Alzheimer's Disease* released details of an animal study showing the benefits of carnosine. In the study, carnosine partially reduced cell death in the brain.[44] And several studies show the brain value of co-enzyme Q10 (CoQ10.) CoQ10 acts as a hydrogen electron shuttle within the mitochondrial membrane (meaning it helps produce energy), and lower levels of CoQ10 are common in patients with PD. A multi-centre, randomized, placebo-controlled, double-blind study performed at the University of California, San Diego found that CoQ10 appears to slow the progressive deterioration of function associated with PD. A report published in 2005 suggests that increased intake of CoQ10 is beneficial in both cardiac disease and PD.[45]

Exercise: A No-Brainer

And if you think you can keep a healthy brain without exercise, you can forget about it (no pun intended)! We've already established that exercise improves metabolism, and is important for circulation. Including aerobic exercise in your day improves oxygen consumption – which your brain needs – and may help to reduce brain cell loss. Try 30 minutes per day of bicycling, gardening, tai chi, yoga or even walking to get your heart pumping oxygen to your noggin. There is also evidence that exercise might improve some symptoms for those who have already been diagnosed with PD.[46]

"Try 30 minutes per day of bicycling, gardening, tai chi, yoga or even walking to get your heart pumping oxygen to your noggin. There is also evidence that exercise might improve some symptoms for those who have already been diagnosed with PD."

You can maintain a healthy brain into your later years. Just think about it (pun intended)!

Brain Protection Nutrient Chart

Nutrient	Suggested dosage
Ultimate Protein Energy Shake™	1 serving twice daily
Ultimate Longevity (with carnosine)™	1-2 capsules/day
Multi-vitamin/mineral complex	Follow label instructions
B vitamins (high-potency)	Follow label instructions
Vitamin C (as mixed ascorbates)	500-2,000 mg/day
Vitamin E (as mixed tocopherols)	800 IU daily
Acetyl-L Carnitine	500 mg 1-3 times/day
Co-Q10	30 mg, 2-3 times/day
Alpha lipoic acid (ALA) R+ form	0 mg, 2-3 times/day
Vitamin B12 (methylcobalamin) sublingual (under the tongue)	1-5 mg/day
X2O mineral sachets	1-2 sachets twice daily in water

SIX

CANCER PREVENTION

The second leading cause of death in North America occurs when cellular mechanism gone awry, in a condition commonly referred to as cancer. In the United States alone, nearly 600,000 people succumb to this disease every year,[47] making it one of the most feared of all diseases. According to the Canadian Cancer Society, an average of 2,865 Canadians are diagnosed with cancer every week. Based on current figures, 38% of Canadian women and 44% of men will develop cancer during their lifetimes.[48] That's a whopping one in three people! Although these statistics are staggering, there is a lot you can do to reduce your risk of developing this disease. The key is simply having the right information and applying it.

Perhaps more than any other, cancer is a disease of the cells. More specifically, it is a disease that results from a problem with cellular metabolism. The normal life cycle of a healthy cell involves a very controlled process of anabolic growth, replication and eventually preprogrammed death (apoptosis). A cancer cell, on the other hand, is one that no longer functions within normal controls and undergoes unrestricted growth. The cancerous cell reproduces uncontrollably, eventually forming a tumor. Through a survival process called angiogenesis, cancer cells emit signals that cause blood vessels to grow from surrounding tissues, eventually connecting the tumors with a fresh blood supply that delivers nutrients for continual tumor growth.[49] The cancer's goal is to survive – even at the expense of the host.

Cells become cancerous through *transformation*, which involves two steps. In the first step, called *initiation*, the cell becomes susceptible to carcinogens. Carcinogens are chemicals, viruses, ultraviolet light, physical traumas, etc., that cause a change in the genetic material of the cell (DNA.) Some forms of hormones have also been linked to cancer. (See Part Two, Condition 12, Her Health: Menopause and Part Two, Condition 19, Prostate Health for more information.) In the second phase of cancer development termed *promotion*, the cell becomes cancerous.

To explain how cancer is able to flourish, it's important to outline two metabolic processes that occur in the body. You will recall that our bodies need oxygen, water, protein, fats and sugars in order to thrive. Our bodies use two methods of metabolizing these nutrients. The first, *aerobic metabolism*, involves the use of oxygen, which is used to burn proteins and fats. As a result, it is also known as oxygen-based metabolism. The second method, *anaerobic metabolism*, does not require oxygen and is involved in burning sugar. While our bodies use both forms of metabolism, anaerobic metabolism is considered more primitive, and is the process used by single-celled organisms like fungi. It is also the preferred choice of cancer.

Dr. Otto Warburg

Early last century, Dr. Otto Warburg observed that tumor cells metabolize glucose anaerobically, even if sufficient oxygen is present. He concluded that the cause of cancer is, in fact, the replacement of oxygen-based metabolism with fermentation of sugar within the cell. In his investigation of the use of oxygen within cells, he identified a family of enzymes called cytochromes that bind molecular oxygen. Known as respiratory enzymes, this group includes iron salts, nicotinamide, pantothenic acid and riboflavin. (The last three are more commonly known as B vitamins.) In fact, Warburg discovered that removing these respiratory enzymes from food inhibits oxygen-based metabolism. Returning these enzymes to the food restores cellular respiration. Warburg also suggested that carcinogens work by interfering with cellular respiration. In1931, Dr. Warburg was awarded the Nobel Peace Prize for his contributions to the understanding of cellular metabolism and its relationship to cancer.

In a lecture to Nobel Laureates in 1966, Dr. Warburg stated that, in order to prevent cancer, we must preserve high blood oxygen levels by maintaining the integrity of the bloodstream (See Part Two, Condition 11, Heart Health) He also recommended keeping a high concentration of hemoglobin (the oxygen-carrying and predominant protein in red blood cells) in the blood, and making sure that the diet contained adequate amounts of respiratory enzymes. Of course, he also advised avoiding known carcinogens, which isn't easy in a world that produces over 600 billion pounds of synthetic chemicals every year[50] – yikes! Warburg's work has become the basis for much cancer research.[51]

Fill'er Up with Oxygen

So, how do we follow Dr. Warburg's advice? First of all, in order to improve blood circulation, we need to make sure arteries are kept clear of debris. Deal with any blood pressure issues immediately. Because the heart is responsible for moving our blood around, you want to protect its health as well. If you haven't already, add regular cardiovascular exercise to your day. Exercise is a proven way to deliver rich oxygen to every cell in your body.

Research shows that exercise is also helpful if you are recovering from cancer. Scientists from Central Washington University in Ellensburg, WA, divided a group of 18 cancer survivors into three groups to compare the effect of low-intensity vs. high-intensity aerobic exercise on various aspects of health compared to a control group. The groups who exercised did so three times per week for a ten-week period. At the conclusion of the study, because researchers saw no significant difference in the results comparing the intensity of the workout for the exercising groups, they combined the results. Compared to the control group, however, researchers found both exercise groups experienced significantly improved aerobic capacity, lower body flexibility, energy levels and quality of life. The exercise group also recorded a significant decrease in body fat. Researchers concluded that exercise is beneficial for people recovering from cancer.[52]

You also want to be sure to feed your body properly. Consider adding a multi-vitamin/mineral supplement to your daily regimen, but be sure that you are consuming bio-available nutrients. You will recall that bioavailable nutrients are the only ones that your body can actually access and use. Some forms of supplemental vitamins and minerals are not in a form that your body can metabolize. Also, learn the best ways to use your foods. The mineral iron, for example, is hard for the body to absorb. Including a source of vitamin C with your iron, however, boosts absorption of the mineral. So, you'll want to have some vitamin C-rich tomato in your iron-filled spinach salad! And be sure to avoid sugars in your diet. Caution: both men and women should consult a qualified health professional before including an iron supplement in their daily health regime.

A review presented in the *Nutrition Journal* in 2004 showed that flax seed and ample portions of fruits and vegetables will lower cancer risk. Cruciferous vegetables like broccoli (a good source of cancer protective indole-3-carbinole) are especially beneficial. The study recommends a diet including adequate sources of selenium (onions and garlic), folic acid (oranges), vitamin B-12 (chicken), vitamin D (fish oils), and chlorophyll (spirulina). Antioxidant carotenoids include beta-carotene (carrots), lycopene (tomato), lutein (fruit) and cryptoxanthin (papaya). Of note, the report also recommends the use of digestive enzymes and probiotics as cancer-preventive nutrients. The study concludes with the bold assertion that following this diet would contribute to at least a 60 to 70 percent decrease in breast, colorectal, and prostate cancers, and a 40 to 50 percent reduction in lung cancer.[53]

It's also a good idea to learn to relax. Researchers have determined that stress and emotional withdrawal might trigger biochemical changes in a cell, which switches the cell over from aerobic metabolism to anaerobic metabolism.[54] Never underestimate the power of a few deep breaths to help diffuse your tension. Finally, and this should go without saying, avoid known carcinogens such as cigarette smoking and over-exposure to the sun.

Cancer Prevention Nutrient Chart

Nutrient	Suggested dosage
Ultimate Protein Energy Shake™	1 serving twice daily
Ultimate Longevity (with carnosine)™	1-2 capsules/day
Multi-vitamin/mineral complex	Follow label instructions
Vitamin C (as mixed ascorbates)	500-2,000 mg/day
Vitamin E (as mixed tocopherols)	800 IU daily
Vitamin D3 (Cholecalciferol)	800 IU daily
Probiotics	Follow instructions
ImmunoCare™	1-2 capsules/day
Melatonin	500 mcg-3 mg prior to sleep
X2O mineral sachets	1-2 sachets twice daily in water

seven

CHRONIC PAIN

According to a recent poll, almost one in five Americans suffers from chronic pain.[55] Although I could not find statistics for Canadians, you can bet the figure is similar in Canada. Chronic pain disables more people than cancer or heart disease, and costs the U.S. economy more than $90 billion per year in medical expenses, disability payments and productivity. Chronic pain interferes with every aspect of a person's life, from work to relationships, to emotional well-being.[56] Unrelenting pain brings the burden of anxiety and depression. A reduced quality of life can lead to thoughts of suicide.[57,58]

Chronic lower back pain is one of the most common musculoskeletal problems, but other common sources of grief include headache, whiplash and pelvic pain, including endometriosis and fibromyalgia. (See Part Two, Condition 9, Fibromyalgia and Chronic Fatigue

Syndrome.) Among other underlying conditions, pain is also a factor in osteoporosis, arthritis and as a result of psychological trauma. In fact, researchers at the University of Western Ontario in Canada found that skin pain and other skin symptoms are common reactions to a stressful event.[59] (See Part Two, Condition 3, Arthritis and Part Two, Condition 4, Bone Health.)

Don't ignore that pain

Your body gives you pain for a reason: it lets you know that something is out of order on the inside of your body where you can't see it. In truth, pain is the main reason that patients go to the doctor.[60] But instead of asking the doctor for a medication to make the pain go away, it's worth your while to figure out the source of the pain – the reason that your body is telling you *through pain* that something is wrong – so that you can possibly correct the underlying issue and relieve pain permanently.

Gender-specific impact of pain

Several studies addressing the impact of pain have found interesting results. It seems, for example, that pain caused an increase in heart rate in 11% of volunteers in a Canadian study – but only in the male participants.[61] Many studies indicate that women experience pain more intensely than do men, but another Canadian study found that sustained low-intensity pain is more distressing to men than to women.[62] Some researchers suggest that the women's sensitivity to pain might explain why females are more likely than males to visit the doctor. The researchers also raised the question of whether the male response to pain was due to gender-role expectations.[63] Guys, whether or not you want to be perceived as a tough guy, you don't want to ignore pain. Remember, it's your body talking to you!

Oh, my aching head!

Another very common source of pain is migraine headache. A migraine is a form of vascular headache that can be sourced all the way to the cells. A migraine occurs when brain cells trigger the nerves to release chemicals that irritate and cause swelling of blood vessels on the surface of the brain. There can be many triggers for migraine, some of which, including barometric pressure, are uncontrollable. Other common triggers include food allergies, flashing lights, sun glare, liver malfunction and fluctuating hormones. Controllable triggers directly related to metabolism include insufficient exercise, too much or too little sleep, and low blood sugar. Low levels of the hormone serotonin have also been associated with migraine.[64] Serotonin acts as a chemical messenger that transmits nerve signals between nerve cells; it also causes blood vessels to narrow. Tryptophan – the precursor to serotonin – is found in extremely high quantities within the alphalactalbumin portion of whey protein.

Studies of migraine prevention show that supplementation of magnesium, coenzyme Q10, riboflavin (B2), and vitamin B12 may help in the prevention of migraine.[65] Magnesium is a muscle relaxant and Coenzyme Q10 helps to improve function of the energy-producing mitochondria of the cell. According to research from the NY Headache Center, New York, regular oral magnesium supplementation may also reduce the frequency of migraine headaches.[66] This is interesting since low levels of magnesium are found in at least 50% of patients during an acute migraine attack.

The B vitamins are important for the central nervous system, and a known effect of riboflavin is to soothe excessive sensitivity of the eyes to light.[67] Regular aerobic exercise might also help to prevent migraine pain.[68] Eating organic foods and avoiding xenoestrogens in your body care products can help you control hormone levels by limiting the input of foreign hormones in your body. (See Chapter 3, The Hormone Connection.) Keeping a migraine journal might help you to isolate triggers for your migraines so that you can limit your exposure to them.

Coping Mechanisms

Current therapies for chronic pain include non-steroidal anti-inflammatory drugs (NSAID) like aspirin and opioids including codeine. Each of them comes with a warning label. NSAID are associated with gastric damage, join deterioration, liver and kidney toxicity and an increase in blood clotting time. (See Chapter 3, Arthritis for more information about NSAID.) Opioids, on the other hand, are addictive and can cause side effects including nausea, respiratory depression, sedation and severe constipation.[69] Over time, opioids lose their effectiveness and, unlike common side effects that pass when the body gets used to a drug, the constipation associated with opioid use lingers.

As you work to address the source of your chronic pain, consider a few of these alternative methods of pain management, including hot packs. In a 2005 study performed in Japan, healthy volunteers with no history of jaw muscle pain were divided into two groups. One group applied hot packs on the masseter muscle (the muscle that raises the lower jaw) and the other used sham packs. Researchers found increased blood flow and improved oxygen levels in the blood of volunteers using the hot packs.[70] Since we know that oxygen is vital for metabolism, and anabolic metabolism is for rebuilding, it makes sense that hot packs would help with the pain and healing of jaw muscle pain – and other muscles as well.

DLPA – Natures Natural Pain Reliever

One natural pain killer that is usually quite effective comes from an essential amino acid called phenylalanine. Amino acids come in two forms, a D and L (right and left) form, which are quite literally mirror images of one another. Most amino acids found in nature are in the L (left) form, therefore phenylalanine normally occurs as L- phenylalanine in foods.[71] Having said this, studies have indicated that D-phenylalanine may offer greater pain relief than L-phenylalanine.[lxxi] Having said this, a mixture of both D and L-phenylalanine (DLPA) has been used to fight pain for nearly 30 years and research seems to indicate that in this combination, the amino acid seems to work best when it comes to fighting pain and inflammation.[72] Aside from its pain-relieving actions, DL-phenylalanine is also a potent antidepressant.[73,74] Although DLPAs exact method of action is not understood, researchers believe it may be due to its suppression of pain-receptive nerves in the spinal column.

Pain-free eating

You already know how to eat to awaken your metabolism. Now you can apply the same understanding (and the same menu!) to coping with pain. A 2005 report in the *Journal of Nutrition* found that the inclusion of eicosapentaenoic acid (EPA), docosahexaenoic acid (DHA), oleic acid, folic acid and vitamins A, B6, D, and E increase the pain-free walking distance in men with peripheral vascular disease (a disease of blood vessels outside the heart.)[75] Other research shows that supplemental vitamin E relieves the pain (and reduces blood loss) of dysmenorrhoea or painful menstruation.[76] And research from the University of South Dakota School of Medicine found that salivary melatonin levels change following acute pain, and concluded that the hormone exerts a pain-relieving effect by increasing the release of the body's natural pain-killing beta-endorphins (these puppies are hundreds of times more powerful than morphine).[77] Avoid stimulants like caffeine that can stress your adrenals. (See Part Two, Condition 1, Adrenal Exhaustion.)

And exercise

Yes, you need to eat a nutritious diet and you have to exercise to reduce the impact of pain in your life. A 2005 German study shows that exercise positively affects individuals with orthopedic disease patterns, including osteoporosis and back pain. Researchers recommended an endurance-training program of 30 to 40 minute duration at least three times per week.[78] Another study showed that lumbar extension exercises – with the pelvis stabilized using special equipment – improves muscle strength, bone mineral density and significant reductions in pain for those experiencing chronic low back pain. The exercise also helps to strengthen the back and possibly prevent lower back injuries. Find an exercise you enjoy, and ease into it. On pain-free days, don't over-exert yourself or you might end up causing yourself more pain.

Pain Prevention Nutrient Chart

Nutrient	Suggested dosage
Ultimate Protein Energy Shake™	1 serving twice daily
Ultimate Anti-Stress™	2-4 capsules prior to sleep
DL-phenylalanine	250-500 mg once or twice daily
Chromium (as LeafBrand)	400-1,000 mcg/day
Multi-vitamin/mineral complex	Follow label instructions
B vitamins (high-potency)	Follow label instructions
Vitamin C (as mixed ascorbates)	500-2,000 mg/day
Vitamin E (as mixed tocopherols)	800 IU daily
Vitamin D3 (Cholecalciferol)	800 IU daily
Melatonin	500 mcg-3 mg prior to sleep

Migraine Chart

Nutrient	Suggested dosage
Ultimate Protein Energy Shake™	1 serving twice daily
Calcium (Glycinate and/or Citrate/Malate)	1,000-1,200 mg/day
Magnesium (Glycinate and/or Citrate/Malate)	1,000 mg/day
Co-Q10	50-100 mg/day

eight

DIABETES

On the cellular level, metabolism is all about energy. The energy, of course, comes from the food that we eat and the way in which that food is transformed into the energy substance ATP. But sometimes there is a breakdown in the metabolic chain of command and systems run amok. A common example of misaligned metabolism is diabetes, and the statistics concerning diabetes are hard to believe. In the US alone, nearly 18.2 million people have diabetes, and over 5 million of them don't know it yet. Sadly, over one million Americans over the age of twenty are diagnosed with diabetes yearly.[79] The prognosis isn't better in Canada. In fact, the Canadian Diabetes Association predicts a 76.5% increase in the diabetes numbers in certain areas of the country by the year 2016.[80] Chances are quite high that you or someone you know is already living with this lifelong metabolic disorder.

Diabetes defined

Diabetes is a metabolic disorder involving the inability to use carbohydrates. You'll remember from our earlier discussion that carbohydrates are broken down into glucose to be used for energy by the trillions of cells in our bodies. (See Chapter 7, Dietary Carbohydrates) Glucose, however, can't enter the cells on its own; it needs an escort. The hormone insulin is that escort. Problems with metabolism occur when, for whatever reason, insulin cannot properly open the cell door for glucose: either no insulin is present, or the cells don't answer when insulin rings the bell. As a result, glucose accumulates in the blood, overflows into the urine, and vacates the body – and the body's main source of fuel is literally flushed down the drain. Without nutrients, cell function is impaired and the high levels of blood glucose can cause damage to blood vessels, eyes, kidneys and nerves.

Types of diabetes

Type 1 diabetes, also known as insulin-dependent or juvenile diabetes, often (but not always) begins in childhood. Type 1 diabetes is an autoimmune condition in which the immune system attacks the insulin-producing cells in the pancreas and destroys them. Because the body does not manufacture sufficient insulin, people with Type I diabetes must take insulin injections in order to survive. Symptoms of type 1 diabetes include increased thirst and urination, constant hunger, weight loss, blurred vision, and extreme fatigue. If not diagnosed and properly treated with insulin, a person with type 1 diabetes can fall into a life-threatening diabetic coma.

Type II diabetes is also referred to as adult-onset-diabetes, although nowadays many teenagers are being diagnosed with the condition. In this version, the pancreas is usually producing enough insulin, and in many cases even too much of the stuff (hyperinsulinemia), but the body cannot use the insulin efficiently. This is known as insulin resistance. Finally, after several years, insulin production decreases. Unlike type 1 diabetes, symptoms of insulin-resistant diabetes develop gradually, and might be due – at least in part – to the biological aging of pancreatic cells. This is why the condition has historically appeared after age 40. With our understanding of metabolism, this explanation makes sense: when cells are not adequately nourished, they become less efficient. (In fact, this might also explain why type 2 diabetes is striking at ever-younger ages. With the high percentage of nutrient-deficient food choices consumed by children and teenagers, cells are not being fed to thrive and they are aging more rapidly. If you are responsible for the care and feeding of a young person, please be sure you are nourishing their cells for optimum metabolic power!)

Symptoms of type 2 diabetes include unusual thirst, blurred vision, frequent urination and slow wound healing. Because some people have no symptoms, type 2 diabetes is a serious threat. Those at high risk for this metabolic disorder have a family history of diabetes, have experienced gestational diabetes (diabetes of pregnancy); or are physically inactive and overweight.

The Lifestyle Factors

As much as diabetes is a chronic health condition with far-reaching consequences, it is also a condition that can virtually be prevented with lifestyle choices, including diet and exercise.[81] In fact, for at least 60 years, scientists have realized there is a relationship between dietary fat consumption and glucose metabolism.[82] A high fat diet impairs glucose metabolism, and the interference has been related to changes in the membrane of the cell. (You will recall that the cell membrane is composed of fat.) Saturated fat, found in fatty meats, cheese and other dairy products, seems to be most damaging to the cell membrane. The best way to protect your cell membranes is to reduce your intake of saturated fats, so choose lean protein sources like chicken, turkey, fish and legumes. For optimum protection, boost your consumption of membrane-protecting omega-3 fatty acids and the preformed omega-6 fatty acid, GLA. You will get your omega-3s from flaxseed, hempseed and especially fish oils and GLA from borage or evening primrose oil.

Reduce your intake of sugary snacks and white foods (white sugar, white pasta, white flour and white rice), and focus your carbohydrate intake on fresh fruits and vegetables to keep your glucose metabolism on cruise control.

You also need to get off the couch. A 2005 report from India shows that obesity, hypertension and insulin resistance increases significantly with a lack of physical activity.[83] Studies also show that exercise promotes metabolism of glucose by skeletal muscle, whether or not insulin resistance is present. Exercising muscles can often take in up to 30 times more glucose than non-exercising ones. This means that people at risk for diabetes – and those who have been diagnosed – should incorporate exercise into their daily lives.[84]

Diabetes is also considered to be an inflammatory condition, and as I wrote about in the chapter about heart health, systemic inflammation is a serious risk factor for cardiovascular disease. (See Part Two, Condition 11, Heart Health) Fortunately, it seems that the regular contraction of muscle during exercise helps to turn off inflammatory mechanisms both within and outside of the skeletal muscle.[85] And if you need more proof of the importance of exercise (whether or not you have been diagnosed with diabetes), a survey of 87 studies of the subject determined that physical activity has a positive effect on fasting blood glucose, total cholesterol, HDL cholesterol and LDL cholesterol in patients with type 2 diabetes. Cholesterol levels, you will recall, are important factors in the health of your heart.[86] Work to include at least 30 minutes of exercise (both cardio and weight resistance), three times per week.

And use opportunities to move your body. Take your dog for a walk, play with your kids or use the stairs instead of the elevator. You have the power to keep diabetes out of your future. The choice is truly yours!

Diabetes prevention Nutrient Chart

Nutrient	Suggested dosage
Ultimate Protein Energy Shake™	1 serving twice daily
FibreLean™ (taken in a full glass of water)	1-3 servings/day
Ultimate Longevity (with carnosine) ™	1-2 capsules/day
Chromium (as LeafBrand)	400-1,000 mcg/day
Multi-vitamin/mineral complex	Follow label instructions
Vitamin C (as mixed ascorbates)	500-2,000 mg/day
Vitamin E (as mixed tocopherols)	800 IU daily

nine

FIBROMYALGIA/CHRONIC FATIGUE SYNDROME

Fibromyalgia Syndrome (FM or FMS) is a painful and chronic disorder affecting up to 3 in 100 Canadians (according to the Canadian Arthritis Society). Women are four times more likely to be affected than men, between the ages of 30 to 60, but most often over the age of 50.

Despite the condition's frequency, diagnosis is difficult. Patients with fibromyalgia usually ache all over, sleep poorly, wake up stiff, are chronically tired and often experience trouble concentrating or have poor memory. Secondary depression and anxiety is also quite common.

If your doctor suspects fibromyalgia, after ruling out other conditions including osteoarthritis, lyme disease and adrenal gland failure (See Part Two, Condition 1, Adrenal Health), he or she will perform a standard test for fibromyalgia by checking 18 distinct sites on the body for pain and tenderness. These sites inlcude: hips, knees, neck, rib cage and shoulders. Fibromyalgia will be diagnosed if at least 11 of these sites reveal deep muscle tenderness. Fibromyalgia is similar to Chronic Fatigue Syndrome (CFS), except that fibromyalgia involves more pain, whereas CFS involves more fatigue. Many people with FMS qualify for the diagnosis of CFS as well.

Chronic Fatigue Syndrome (CFS) is a condition of metabolic exhaustion, which involves severe fatigue after even the smallest degrees of physical or mental exertion. Increased activity typically causes symptoms to worsen, and over-doing it can lead to days of recovery. The US Center for Disease Control has established working criteria for the diagnosis of CFS, including persistent or relapsing debilitating fatigue lasting six months or more that does not improve substantially with rest. In addition, four of the following criteria must be present:[87]

- impaired concentration or memory
- headaches
- malaise after exertion
- muscle pain
- multi-joint pain
- non-restorative sleep
- painful or tender nodes
- sore throat

Science has not yet uncovered the cause of FMS or CFS, but theories abound. Researchers have found elevated levels of a nerve chemical signal, called substance P, and nerve growth factor in the spinal fluid of FMS patients.[88] Serotonin levels have also been shown to be low in patients with fibromyalgia.[89] Serotonin is a neurotransmitter or chemical messenger that transmits signals between nerve cells as well as causing blood vessels to narrow. Changes in serotonin levels can alter mood, emotion, sleep and appetite. Other studies of fibromyalgia suggest that the central nervous system is somehow supersensitive, causing a disturbance in the perception of pain. Fibromyalgia does not involve tissue inflammation, yet some evidence suggests the involvement of the immune system.[90,91] Research presented in the journal *Medical Hypothesis* suggests that FM is the result of resistance to thyroid hormone, indicating a metabolic basis for the condition.[92]

Like fibromyalgia, the exact cause of CFS in unknown. Some research suggests that the Epstein Barr Virus is involved, yet evidence is inconclusive. Other theories link CFS to anemia, arthritis, hypothyroidism and intestinal parasites. Candida overgrowth might also be a factor. (See Part Two, Condition 10, Gut Health.) A study published in 2004 found an increase in apoptosis (programmed cell death) of key immune system cells, in a fashion consistent to an underlying viral or toxic illness. Researchers concluded that patients with CFS have an underlying abnormality in their immune cells.[93]

"Drink at least 10 eight ounce glasses of properly filtered water daily. For best effects, try adding an X_2O water sachet to your water twice daily in order to keep your body as alkaline as possible. Water helps to flush out toxins and helps to relieve muscle pain."

Stay Strong

As with any other health condition, prevention is key. Keeping the immune system in working order could help to prevent FMS, as well as fight an invading virus that might lead to CFS. You accomplish this, of course, with a nourishing diet high in antioxidants, regular exercise and adequate sleep nightly. Reducing stress is also a wise goal to preserve your health and prevent FMS and CFS. Research shows disturbed adrenal gland function in adolescents with CFS,[94] which is a typical consequence of chronic stress. (See Part Two, Condition 1, Adrenal Health.)

Coping Mechanisms

If you suspect a diagnosis of CFS or FMS (or already know for sure), there is much you can do to improve the way you feel. First off, be sure that you follow the metabolism-boosting eating plan described in Chapter 8. And despite your energy level, you will benefit from regular, moderate exercise. Several studies confirm that aerobic exercise is beneficial to patients with FM and CFS. Drink at least 10 eight ounce glasses of properly filtered water daily. For best effects, try adding an X_2O water sachet to your water twice daily in order to keep your body as alkaline as possible (see chart below and Appendix I for recommendations). Water helps to flush out toxins and helps to relieve muscle pain. Add a few mugs of green tea as well, to reduce the impact of stress. Indian researchers published a study in the *Journal of Medicinal Food* in 2005 indicating that green tea extract could be used as to ameliorate oxidative stress in the management of CFS.[95]

Don't forget about your serotonin levels if you suffer with fibromyalgia. Supplementing with 5-hydroxytryptophan (5-HTP) or with prescription L-Tryptophan has been shown to significantly improve symptoms of depression, anxiety, insomnia and pain associated with FMS.[96]

One of the body's most powerful antioxidants is called glutathione (GSH). GSH is also essential to the health of your immune system. It should not come as any surprise that GSH levels are often very low in people with CFS, especially in muscle cells. Researchers from McGill University in Montreal, Quebec believe that certain immune cells may actually steal scarce GSH precursors from the energy starved muscle cells in order to maintain a certain degree of immunity. Since GSH is also essential to aerobic muscular contraction, the muscle cells become deprived of their vital energy substance and CFS ensues.[97]

Studies have found that properly processed (low heat) whey proteins offer a superior advantage when it comes to elevating the body's GSH levels.[98] The reason for this is they contain high levels of the building blocks of GSH, with the highest GSH builders being found in the alpha portion of the whey. Whey protein is so effective at raising GSH levels, it has been used to help treat immune compromised diseases like cancer[99] and cystic fibrosis.[100] Since athletes can also experience reductions in their GSH levels after bouts of intense exercise, whey isolates can also be used to help return their GSH to healthy status.[101]

The Fantastic Four

L-Carnitine

You'll recall that cellular metabolism is controlled to a large degree by the cells power plants, mitochondria. Fatty acids happen to be the primary fuel source for energy production. However, they are not able to cross the mitochondrial membrane without help. This help comes in the form of a natural vitamin-like substance, L-carnitine, which acts as an energy transporter by shuttling fatty acids directly into the mitochondria to be burned as energy.[102]

Since CFS is believed to be caused by a defect in mitochondrial function, and past research has revealed low carnitine levels in CFS patients, researchers have often wondered if carnitine may play an essential role in its treatment. Researchers from the Chronic Fatigue Syndrome Center in Chicago decided to find out. Researchers compared the effects of L-carnitine against the well known fatigue-fighting drug Amantadine in 30 CFS patients. Both L-carnitine and Amantadine were administered alternately over a two month period with two week rest periods in between. At the end of the study, the researchers were shocked to find that 29 of the 30 test subjects showed statistically significant clinical improvement, whereas the drug treatment was poorly tolerated by all test subjects, with half of them stopping the medication due to side effects.[103]

NADH

NADH (nicotinamide adenine dinucleotide) is an activated form of the B vitamin niacin, which is essential to any living cell's energy production. In fact, NADH is the coenzyme that plays the primary role in the release of energy within the mitochondria. Aside from this, it is also a powerful antioxidant.

Researchers from Georgetown University School of Medicine, Washington, D.C., have indicated that NADH may be a valuable adjunctive therapy in the management of the CFS.[104] Researchers from San Juan, Puerto Rico discovered that CFS patients who received NADH over a 24 month period reported dramatic and statistically significant reduction in their symptoms.[105] Due to the fact that NADH is so easily degraded by stomach acid, only supplement with enteric-coated NADH.

Co-Q10

Another nutrient essential to cellular metabolism is co-enzyme Q10 (co-Q10). Although mostly known for its cardiovascular benefits, Co-Q10 is essential to creating the spark that ignites fatty acid oxidation (burning).[106] But Co-Q10 also acts as a powerful antioxidant inside the mitochondria (where the majority of free radicals are formed).[107]

ALA

And the last – but certainly not least – cellular metabolism nutrient is alpha lipoic acid (ALA). Aside from being one of the most powerful and versatile of the antioxidants (it is both water and fat soluble), ALA helps to enhance energy levels.[108] Research from the Linus Pauling Institute indicates that ALA in association with carnitine can greatly increase energy production and reverse the negative energy decline associated with aging in animals.[109]

Although there is no known cure for FMS and CFS, it is possible to reduce – and in some cases eliminate – the impact that these conditions have on your life. Remember the power of your metabolism when it comes to providing your cells with abundant energy!

FMS and CFS Prevention Nutrient Chart

Nutrient	Suggested dosage
Ultimate Protein Energy Shake™	1 serving twice daily
Ultimate Anti-Stress™	2-4 capsules prior to sleep
Ultimate Longevity (with carnosine) ™	1 capsule/day
Multi-vitamin/mineral complex	Follow label instructions
B vitamins (high-potency)	Follow label instructions
Vitamin C (as mixed ascorbates)	500-2,000 mg/day
Vitamin E (as mixed tocopherols)	800 IU daily
Co-Q10	50 mg, 2-3 times/day
L Carnitine	500 mg 1-3 times/day
Alpha lipoic acid (ALA) R+ form	50 mg, 2-3 times/day
NADH (enteric coated)	5 mg/day
X2O mineral sachets	1-2 sachets twice daily in water
Magnesium	600-1,00 mg/day

ten

GUT HEALTH: Acid Reflux/Heartburn, Candidiasis, IBS, Crohn's disease, Colitis and Celiac disease

At the risk of sounding like a broken record, I want to remind you again that metabolism involves changing energy to a usable form at the cellular level and, of course, our energy comes from food. This means we don't just feed the stomach when we eat, but instead our meals are broken down into the various fuels required by every one of the trillions of cells in our bodies. And in order to get the best metabolic bang for our breakfast, lunch and dinner, we need to have a healthy digestive tract. But if you watch television or flip through your favourite magazine, it won't be long before you see an advertisement for antacid tablets or remedies for constipation. And what are disorders like heart-burn or constipation if not digestive problems? If we have digestive troubles, how can our bodies fully benefit from our foods? In fact, many North Americans are literally starving to death on full stomachs due to the nutrient-void foods they are constantly placing in their systems.

Let's talk about those products for a moment. While bowel aids might get things moving for you, and a tablet might temporarily take the sting out of your meal, have these products really solved the problem in the end? No! If they did, it would be unlikely that 60 to 70 million people in the US alone would be suffering with digestive complications.[110] So, instead of using products that might help with our short-term discomfort when we suffer digestive backlash, it's more proactive and pro-metabolism to address the underlying factors that got us to that place to begin with, not to mention the ones that are still keeping us there.

A bit about Digestion

In order to deliver the hidden nutrients in food to the trillions of cells throughout the body, we must first break apart and release the nutrients through digestion. Our bodies use both mechanical and chemical tools to achieve this. Mechanical devices include chewing and the churning of our stomachs. The chemical component includes such things as gastric juices (also known as stomach acid), bacteria and digestive enzymes. There are four major digestive enzymes: protease to break down protein, lipase to break down fat, lactase to break down dairy and amylase to break down carbohydrates.

After passing through the stomach, bile from the gall bladder emulsifies the fats in our food, and the concoction passes through the small intestine. Tiny vacuum-like structures in the small intestine called villi start to absorb the nutrients. As the food works its way to the large intestine, nutrients have been removed and the process of preparing waste for disposal takes place. Beneficial bacteria go to task with final break-down duties and water is absorbed into the body. Waste products are stored in the colon until the final exit from the body – often referred to as "the big flush!" (Unless your lavatory is on the fritz that is).

The process seems simple enough, but only if we support it. First of all, many of us rush through our meals and end up swallowing very large particles of food. While stomach acid is quite strong, it works much more efficiently when given smaller bits to work on. Be sure to chew every mouthful until it is liquid. In fact, the saliva in your mouth contains amylase, and if you chew your food long enough, carbohydrates can be almost completely digested before you swallow!

Secondly, the best source of vital enzymes is actually our food. But not every food qualifies. Only uncooked, unfrozen, unprepared foods like fresh fruits and vegetables supply us with enzymes. Prepared foods like cereal, bread, pasta, pizza and French fries are devoid of enzymes and are considered "dead foods." Our bodies must take from enzyme stores to digest these processed foods, and over time, we run an enzyme deficiency. An oversupply of processed foods, along with red meats, can drain the potency of our stomach acid as well, leading to weakened digestion. Weakened digestion means that your body isn't benefiting from the nutrients within the foods you eat. Say it with me now: if your body isn't benefiting from the foods you eat, your metabolism is going to suffer.

How is your Digestion?

The incomplete digestion of food shows itself in myriad ways. Common symptoms include the aforementioned heart-burn and constipation, but also include such conditions as *Candidiasis*, Crohn's disease, celiac disease, and colitis. Even seemingly unrelated conditions – including eczema and psoriasis – can often be traced back to inefficient digestion. (See Part Two, Condition 20, Skin Health.)

Acid Reflux/Heartburn:

After we eat, our food travels down the esophagus, a tube that is supposed to flow one-way as it connects the mouth and stomach. Normally, a group of muscles called the esophageal sphincter squeeze shut to prevent the upward migration of stomach acid, but once in a while, a malfunction occurs. When stomach acid backs up into the esophagus, the result is a painful irritation called heartburn. Heartburn that occurs more than twice a week is considered gastroesophageal reflux disease or GERD. Left untreated, GERD can lead to scarring of the esophagus, leakage of stomach acid into the lungs, and a pre-cancerous condition called *Barrett's esophagus.* Because acid indigestion often mimics the early stages of angina and heart disease, heartburn should be taken very seriously.[111]

If you frequently suffer from heartburn, seek the advice of your health care provider. GERD is often associated with pregnancy, cigarette smoking, overweight and alcohol consumption. To reduce occurrence of acid reflux, avoid caffeine including chocolate, citrus fruits and tomato-based foods, fatty and fried foods, spicy foods, peppermint and other mints, garlic and onions. Focus your diet on fresh, raw vegetables and lean protein. Eat approximately five smaller meals each day and make sure to chew your food very well.

Candidiasis

While many people haven't heard of *Candidiasis* or candida, it is a common problem that originates in the digestive tract. You will recall earlier in the discussion of how our digestive system works, how we rely on certain types of bacteria for assistance. The large intestine is host to trillions of bacteria: some help us out, while some hitch along for a free ride. Our beneficial bacteria don't let the freeloaders get out of hand, and are quite able to control their numbers. So, when we are healthy, these freeloaders don't cause us much trouble. When our system is out of whack, however, the freeloaders can quickly become unwanted guests. Candida bacteria are freeloaders that easily overstay their welcome.

Let's say, for example, you drink chlorinated water or have taken antibiotics to treat some infection (and who hasn't in their past?). These bacteria-killers are non-discriminating, which means they target and kill all bacteria in our bodies, including the ones we can't afford to live without. That would be okay if the good guys reproduced as fast or faster than the freeloaders like candida, but sadly, the opposite is true. In no time after a round of antibiotics, the bacterial balance shifts to the favour of candida. After a while, candida bacteria begins to morph. They grow tiny feet-like projections that cling to the walls of the intestines, eventually creating tiny holes in the intestinal lining. The candida makes an escape and travels through the bloodstream to tissues throughout the body. Then, they form colonies that reproduce like wildfire.

And bacteria is not the only things that escape through the holes in the intestine. Tiny undigested food particles can also leak through these holes and travel through the bloodstream. Once detected, the immune system attacks these particles, leading to an allergic reaction. Have you noticed the increase in food allergies lately? Have you experienced any yourself? Other symptoms of candida overgrowth include vaginal infection (particularly recurring events), acne and eczema, cold hands and feet, white coating on tongue, constipation or diarrhea, depression and mood swings, muscle aches, brain fog, pre-menstrual syndrome (PMS), and dandruff.

The best way to eliminate candida is to boost your digestive processes. Eat enzyme-rich, high-fibre foods. Small, frequent meals are easier to digest, leaving little for the candida to dine on when food reaches the large intestine. Supplement your diet with organic fruit and vegetable fibres and probiotics – a fancy term for the good bacteria that live in our bodies. Acidophilus and bifidus are common probiotics. Consider taking fructooligosaccharides (FOS) from natural fibres like inulin, the preferred food of our beneficial bacteria. And be sure that your diet is high in fibre in order to help clear the candida from your body.

Irritable Bowel Syndrome (IBS)

At least 22 million Americans and approximately 10 to 20% of adults – especially women – worldwide suffer from Irritable Bowel Syndrome (IBS). The National Institute for Diabetes, Digestive and Kidney Disease (NIDDK), predicts that IBS causes 34,000 hospitalizations, 3.5 million physician office visits, and 2.2 million prescriptions annually.[112] With symptoms including abdominal cramps and bloating, flatulence and belching, feeling of urgency or incomplete evacuation, abnormal stool frequency, diarrhea or constipation, or diarrhea alternating with constipation, IBS can impact all aspects of life including sleep, employment, leisure time, travel and sexual functioning. It is also associated with anxiety and depression. In fact, research found depressive disorders in 94% of patients with IBS. The bowel condition is also a factor for 49% of patients with fibromyalgia, 51% of patients with chronic fatigue syndrome (See Part Two, Condition 9, Fibromyalgia and Chronic Fatigue Syndrome) and 50% of patients with chronic pelvic pain.[113]

Although there is a genetic component to this condition, diet and everyday stress play a major role in IBS. Candida infection can be a factor, as can celiac disease. Celiac disease (also known as gluten allergy or gluten intolerance) involves an allergy to gluten, a protein found in wheat and other grains. Medical tests can pinpoint whether celiac disease is a cause for IBS. If gluten is found to be the culprit, all foods containing gluten, including wheat, rye, barley and often oats must be avoided.

Many people rely on traditional medicines to relieve their IBS symptoms. However, there is quite a bit of buzz coming from Europe on the effectiveness of one natural therapy that consists of a combination of peppermint and caraway oil. In a clinical study performed in Germany, 45 patients with non-ulcer dyspepsia and the majority with IBS used a combination of peppermint/caraway oil in enteric-coated form. The trial was double-blind and placebo-controlled (the most respected of scientific studies). After only two weeks of using the peppermint and caraway oil capsule, almost one half of the patients (42.1%) in the test group were completely free pain-free. By the fourth week of treatment, 63% of patients were pain-free.[114] (For more information on this unique combination, please see Appendix I).

You can also reduce the impact of stress on your IBS by taking advantage of massage, meditation and walking.

Crohn's disease

Crohn's disease often involves the inflammation of the intestinal wall, but can occur anywhere along the digest tract, including the mouth. Inflammation and obstructions lead to chronic pain and malnutrition. Symptoms include diarrhea, abdominal pain, fever, loss of appetite and weight loss. There is often inflammation of the joints, skin and eyes. Rectal bleeding frequently occurs if the inflammation is in the large intestine, increasing the risk of cancer after many years. The disease has become much more common in Western and developing countries in the last several years.

Along with inflammation of the intestinal tract, those with Crohn's disease suffer inflammation of the spine, pelvic joints and the inside of the eye. Crohn's disease is linked to other digestive disorders, including gallstones and inadequate absorption of nutrients. There is no cure for Crohn's disease, so the most important goal is to control inflammation and address nutrient deficiencies.[115]

Celiac Disease

For millions of people worldwide, eating a meal containing gluten results in damage to the small intestine. Gluten is a protein found in all types of wheat and the related grains, rye, barley, spelt and kamut. An offending food triggers an immune system attack involving abdominal cramping and gas, chronic diarrhea or constipation (or both), steatorrhea (fatty stool), and anemia. The allergic reaction damages the villi or tiny vacuums lining the small intestine that supply our blood with nutrients. As a result, those with Celiac disease are at serious risk of the nutrient deficiencies that lead to metabolic weaknesses.

Obviously, avoiding gluten is not only extremely important to someone with this disease, it is imperative. Supplementing with essential fatty acids will help to nourish the villi in the intestines. Probiotics are useful to keep the optimum balance of bacteria in your intestines.[116]

Colitis

Another digestive condition, colitis involves inflammation and ulceration of the mucous membranes of the large intestines.

Symptoms include bloating, gas, pain, bloody diarrhea and fever. Colitis often involves hard stools, which create extra work for the colon muscles when it is time for elimination. Walls of the colon then bulge outward in little pouches called diverticula. The opening of the diverticula can bleed into the intestine and fecal matter can become impacted in the diverticula, causing inflammation and infection. Similar to Crohn's disease, inflammation can occur outside the intestines for those suffering with colitis. Other points of inflammation include joint inflammation (arthritis), along with inflammation of the spine, liver, bile ducts and the inside of the eye. Long-term colitis creates an increased risk of colon cancer.

Although the exact cause of colitis is unknown, poor eating habits, chronic stress and food allergies are contributing factors to this condition. Fortunately, these are all factors that you can easily control. Isolate food allergies and eliminate offending foods from your diet. Eliminate dairy foods as well. Focus your diet on fruits and vegetable and lean sources of protein. Avoid raw foods that can irritate the lining of the intestine.

If you use antibiotics in the management of colitis, be sure to follow up with probiotic supplementation to restore balance in the intestines. Supplemental digestive enzymes will help with food digestion, and essential fatty acids will help to reduce inflammation. Vitamin A is particularly important for healing mucous membranes.[117,118]

Anti-inflammatory fats

As you are now aware, many gut related disorders have a definite inflammatory component to them. Specialized essential fats (mentioned below) have been documented to exert powerful anti-inflammatory reactions where intestinal problems arise.[119] The predominant helpers in this area include omega-3 fats such as cold-water fish, krill oil, flaxseed, hempseed and walnuts, as well as gamma linolenic acid or GLA (from borage or primrose oil) and conjugated linoleic acid or CLA. These fats not only work to block powerful inflammatory messengers, they also block the production of Substance P, a chemical produced in the brain that promotes pain and inflammation.[120] Eicosapentaenoic acid (EPA) and docosahexaenoic acid (DHA) seem to be the predominant forces behind the anti-inflammatory capabilities of both omega-3 and omega-6 essential fatty acids. EPA and DHA are exclusive metabolites of omega-3 essential fats, found in abundance in fish oil. (See Chapter 6 for more information.)

What does it mean, metabolically-speaking?

If you suffer from any of these digestive conditions, you need to consider your body as a whole. Not only must you find relief from your intestinal symptoms through proper diet and lifestyle choices, but you must also promote healing by juicing up your metabolism. Only when every cell in your body is benefiting from the right nutrients can you expect more complete healing.

Keeping the digestive system strong is an important first step in boosting your metabolic power. When possible, eat a variety of fresh, raw, enzyme-packed fruits and vegetables. Include sources of lean protein, as well as nuts and seeds to increase your dietary intake of essential fatty acids.

Gut Health Protection Nutrient Chart

Nutrient	Suggested dosage
FibreLean™ (taken in a full glass of water)	1-3 servings/day
LaniFlex (with SierraSil®)	Follow instructions
Probiotics	Follow instructions
L-glutamine	1-3 tsps./day
IBS Regimint™	Follow instructions

eleven

HEART HEALTH

Throughout this book, I have focused on metabolism at the cellular level and the various disorders that result from the insufficiency of that metabolism. It is important to realize that oxygen is the number one metabolic factor (next to water) because it is the driving force behind metabolic combustion. Having said this, the health of your heart – pumping blood through 60 thousand miles of blood vessels each day to every cell throughout your body – is essential to the delivery of this oxygen. Yet, very few people ever realize just how unhealthy their heart really is. That is, until it's too late!

In 2001, over 64 million Americans were living with one or more forms of cardiovascular disease, including heart disease and stroke. Americans are far from alone. According to the Canadian Heart and Stroke Foundation, over one third of deaths in 1999 were due to cardiovascular disease (CVD), and eighty percent of Canadians have one or more risk factors for heart disease. Because the disease affects so many people, the cost of CVD in Canada topped $18.5 billion in 1998 alone. While in the past, heart disease has been considered a man's disease, new figures show that risk of heart disease becomes virtually equal between men and women as we age.[121,122] Heart disease is called a silent killer, because often the first symptom is a major cardiac event. Other symptoms include sudden weight gain (three pounds in a day or five pounds in a week), swelling in the legs or ankles, pain or swelling in the abdomen, shortness of breath unrelated to exercise, fatigue, pain or numbness in the arms, back, neck or chest, and, of course, chest pain.

Who's at risk?

Known risk factors include family history, a highly-processed diet including too much sugar and too many unhealthy fats, sedentary lifestyle, and cigarette smoking. It is also well known that hyperten- sion (high blood pressure), diabetes and obesity are linked to heart disease. There isn't room in this book to go into great detail about all of the metabolic factors that influence heart health, but understanding a few hidden contributors to CVD just might save your life.

C-Reactive Protein

Do you remember the last time you bumped into the counter or cut your finger? Instantly, your body went into protect and repair mode. You experienced swelling, heat, redness and pain while your immune system set to work. This combination of mechanisms for isolating an injury and setting the stage for repair is called inflammation, and it's the only defense your immune system has to protect you from threats. And while it is evident that this defense system is in place when you bump your shin, you might not be aware that it also takes place inside your body. An obvious example of internal inflammation is arthritis. (See Part Two, Condition 3, Arthritis.) But, as in the case of arthritis and many other health conditions, inflammation can get out of control and lead to serious problems elsewhere in the body. This is known as systemic inflammation.

It makes sense that we can have inflammation in our bodies without being aware of it. Unfortunately, however, the presence of inflammation always indicates that our immune system is at work protecting us from some threat, whether we are aware of that threat or not. And persistent inflammation is not healthy for our cells. One way to measure the amount of inflammation in our bodies is by checking serum C-reactive protein (CRP) levels. And many studies show that the risk of heart disease increases with the levels of CRP found in the blood.

According to the American Heart Association, studies of men, women and the elderly show that the risk for heart attack in people with CRP levels in the upper third is twice as high as those whose CRP levels measured in the lower third. Some data shows that elevated CRP is associated with the functioning of our arteries,[123] which contributes to the health of the heart because constricted arteries cause the heart muscle to work harder. Insulin-resistance and the metabolic syndrome (See Part Two, Condition 17, Metabolic Syndrome.) are also associated with elevated CRP levels. Be sure that your doctor includes a CRP test with your regular physical exam.

Homocysteine

Homocysteine is a natural by-product of various biochemical actions in our bodies, and is normally metabolized to the essential amino acid methionine. Unfortunately, aging and chronic health conditions (including inflammation) hinder our ability to make the conversion, and blood homocysteine levels rise. Elevated homocysteine levels are associated with Alzheimer's disease, dementia, kidney failure and heart disease to name but a few. [124,125,126] A 2005 study published in the prestigious *American Journal of Clinical Nutrition* shows that hyperhomocysteinemia (elevatated homo-cysteine) can also contribute to a pro-inflammatory state in the arteries. This might be one explanation of the link between homocysteine and heart disease.[127]

If you are starting to think that lowering homocysteine levels is good for your health, you are on the right track. In observational studies, a 25% lower homocysteine level is linked with approximately 20% less stroke and 10% less coronary heart disease.[128] Luckily, it is very easy to reverse high homocysteine levels by making a few simple dietary changes. Researchers have found that high homocysteine levels are associated with deficiencies in folic acid and the B vitamins, particularly B-6 and B-12. Knowing what you do about metabolism, this should make sense when we consider that, as we age, our ability to absorb the B vitamins through the stomach becomes severely hampered. (This explains why many of the elderly require regular injection of vitamin B-12.) To help keep your homocysteine levels under control, be sure to get adequate folic acid along with vitamins B-6 and B-12.[129] If you are concerned that you aren't getting these nutrients in your diet, take them in supplement form. Researchers have also found that drinking coffee and tea increases homocysteine concentrations by up to 20%.[130] Consider reducing your intake of these beverages if you know that your homocysteine levels are a concern. If you don't know about your homocysteine levels, have your doctor check it along with your CRP level at your next physical.

Cholesterol

With the bad rap cholesterol gets, you'd think it was one of the worst things for your health. But that's only half the story. Believe it or not, cholesterol is the beginning substance for your sex hormones, so you never want levels that are too low. We have two kinds of cholesterol: High Density Lipoprotein (HDL) and Low Density Lipoprotein (LDL.) LDL is the troublesome one because, once oxidized, it causes a buildup of fat and cholesterol in your arteries. Heart attack or stroke can occur if arteries become clogged. While you're sitting at the doctor's office with your sleeve rolled up to learn your CRP and homocysteine levels, you can also learn about your blood cholesterol. You want your LDL count to be less than 130. To help keep LDL levels in a healthy range, avoid saturated fats like dairy and fatty cuts of beef and pork, as well as insulin stimulating high glycemic foods. Insulin has also been shown to raise cholesterol levels. For lean sources of metabolism-boosting protein, eat chicken, turkey, fish and legumes. Replace fatty sweets like ice cream with lots of fresh fruits.

HDL, on the other hand is your friend. One of its roles is to remove cholesterol from the body, which is one reason you want to keep your HDL levels high. Regular exercise is a great way to keep HDL levels up. It is best to strive for HDL levels higher than 35. While monitoring your LDL is important, you should also track your total cholesterol level to make sure the whole cholesterol story is healthy. As simple as it sounds, this is accomplished by adding your LDL and your HDL level together. A normal total cholesterol level is less than 200.

Other factors that influence heart health

How you are feeling physically, it seems, has a lot to do with how you are feeling emotionally. A 2005 study in France showed that, on average, men who were experiencing a depressed mood had 46% higher CRP and increased levels of other inflammatory markers than controls. Researchers concluded that depressed mood is associated with heart disease.[131] For many reasons, depression should not be taken lightly.

Other research suggests that the quality of the air we breathe impacts our heart health. A study performed at the Division of Toxicology at Tulane University School of Medicine showed that for every 10-microg/m(3) increase in fine particulate matter in the air, there appears to be a 2.1% increase in the number of deaths related to ischemic heart disease. Not surprisingly, particulate matter has been linked to increased levels of systemic inflammation biomarkers such as CRP.[132] If you live in a high pollution area, do your heart a favour and consider purchasing an air filter for your home.

And now the good news!

At this point, it shouldn't shock you to learn that properly nourishing your metabolism is also very good for your heart. Focus your diet on fresh vegetables and fruit, lean sources of protein, nuts and seeds, whole grains and plenty of fresh, filtered water every day. And be sure to eat fish and/or supplement with properly formulated essential fatty acids (see Appendix I for recommendations). A 2005 study performed in Athens, Greece measured the impact of dietary fish on heart health. Nearly 3,000 participants were evaluated using a food questionnaire. Compared to non-fish consumers, participants who ate more than 300g of fish per week had an average 33% lower CRP levels in their blood. Other inflammatory markers, such as interleukin-6, were also substantially lower. Eating less than 300g of fish per week also showed significant improvement over non-fish eaters. Researchers concluded that eating fish is independently associated with decreased levels of pro-inflammatory chemicals in the blood.[133]

Pump it up!

You've known for years that aerobic exercise is good for your heart, and now you understand the impact of inflammation on your health. So, you're ready to learn another explanation of why: while exercise temporarily raises inflammatory markers in the bloodstream, in the long run, those who exercise regularly benefit from a long-term anti-inflammatory effect.[134] A 2005 study from the Department of Internal Medicine at the University of Connecticut tested the impact of aerobic exercise on pro-inflammatory markers.

Twenty-eight coronary heart disease patients exercised at 70 to 80% individual maximal heart rate for 12 weeks. At the end of the program, diabetics measured an average 40.5% decrease in CRP levels. In the study overall, there was a 46% reduction in the number of subjects in the high risk category. Researchers concluded that aerobic exercise is an effective method of reducing CRP and other pro-inflammatory markers.[135] The next time you think about your metabolism, think of your heart and get moving!

Heart Health Nutrient Chart

Nutrient	Suggested dosage
Multi-vitamin/mineral complex	Follow label instructions
B vitamins (high-potency)	Follow label instructions
Vitamin C (as mixed ascorbates)	500-2,000 mg/day
Vitamin E (as mixed tocopherols)	800 IU daily
Co-Q10	50 mg, 2-3 times/day
Vitamin K2	15 mg, 1-3 times/day

twelve

HER HEALTH: Menopause

Despite what you might be led to believe (or experience yourself!) menopause is not a disease. Instead, the end of menstruation simply marks the passage into a new phase of a woman's life (but try telling the majority of women throughout North America that!). The ovaries have stopped producing eggs, estrogen levels drop and there is no possibility of pregnancy – yahoo! And because the average human life span has increased, women are spending increasingly more of their lives in a hypoestrogenic (low estrogen) state. Although a completely normal experience, menopause is not however, without the potential for discomfort and can exacerbate risk factors for chronic conditions like heart disease and osteoporosis.

Menopause officially begins one year after the last menstrual period, and the time leading up to the complete cessation of menstruation (about three to six years) is known as the climacteric or peri-menopause. Peri-menopause can begin as early as age 35, but usually begins in the mid-to-late forties. Average age of natural menopause in North America is 51 years, but medication, surgery, radiation and autoimmune conditions that affect the reproductive system can trigger menopause earlier.

As ovaries wind down production of eggs during peri-menopause, hormone levels swing as wildly as they did during puberty. Menopause finally occurs when the ovaries lose sensitivity to the luteinizing hormone that triggers the release of an egg, and the menstrual cycles stop. Hormonal shifts trigger changes in the body that can aggravate other existing hormonal conditions. Consider how many of the following symptoms, normally linked with menopause, are similar to symptoms associated with other conditions. Anxiety attacks, increased blood pressure, bone pain, low energy, heart palpitations, insomnia or restless sleep and mood changes may be due to compromised adrenal gland function. (See Part Two, Condition 1, Adrenal Exhaustion) Undiagnosed clinical or functional hypo-thyroidism might also be a factor in some symptoms, including rapid onset of wrinkles, bloating, gas and indigestion, low energy, sleep disturbances, joint and muscle pain and painful intercourse or diminished sex drive. (See Part Two, Condition 22, Thyroid Health.) Liver and colon congestion can also complicate the metabolism of estrogen, leading to an increase in symptoms. (See Part Two, Condition 14, Liver Health.)

Symptoms

Over 80% of women experience some symptoms at menopause, and more than 25% of women will seek medical attention because symptoms are affecting their quality of life.[136] The most obvious symptom is the cessation of menses. Other common symptoms include anxiety, hot flashes and night sweats, problems with sleep, vaginal itching (with or without discharge), brain fog and memory problems, mood swings, irritability and depression, migraine headaches, new environmental allergies and food sensitivities,weight gain, urinary incontinence (aggravated by coughing, sneezing or laughing) and urinary tract or vaginal infections. Factors such as use of birth control pills, stress, diet and lifestyle will play a role in the age of onset and severity of symptoms.

The Good & Bad Estrogen

As women (and men) age, they can often experience the debilitating effects of what is now referred to as estrogen-dominance, a condition in which healthful estrogens (2-hydroxy estrone) are pushed out of the way by unhealthy estrogens (16-hydroxy estrone). This is believed to be at least in part due to the multitude of xeno-estrogens (estrogen mimicking compounds) disrupters we subject ourselves to, as well as a decline in many of the hormones that help balance estrogens in the body. This disruptive hormone imbalance is responsible for many of the conditions women attribute to peri-menopause, PMS and menopause, including weight gain – especially in the abdominal area – and increased breast cancer risk.[137]

Now for the good news. Through diet, exercise and proven nutrient supplementation, both women and men can alter the way natural estrogens, xenoestrogens and their co-activators react with the various cells of the body, ultimately decreasing their cancer risk and fat storing ability. In other words, with the right nutrients, you can actually turn the right switches on and keep the wrong ones off!

Out of all the incredible nutrient discoveries, I3C seems to have the greatest ability to convert harmful estrogens into positive ones. In fact, research from the Strang Cancer Research Laboratory, Rockefeller University and The Institute for Hormone Research (all in New York) shows that I3C not only has the ability to create beneficial estrogens in the body, but it also exhibits the ability to block harmful estrogens from activating their receptors, thereby blocking the undesirable estrogenic messages to the body.[138,139]

Treatment Options: Hormone Replacement Therapy

Introduced about 70 years ago, (estrogen) hormone replacement therapy (HRT) was once considered top-of-the-line treatment for menopausal symptoms. But the Women's Health Initiative Study cast a shadow over HRT as a panacea for menopausal complaints. The study, which was designed to clarify the risks and rewards of combination HRT, was halted early because researchers determined that the combination of estrogen and progestin (synthetic progesterone) increased the risk of invasive breast cancer, heart disease, stroke and pulmonary embolism. A companion paper also determined that estrogen therapy, used alone for 10 years or more, greatly increased the risk of ovarian cancer.[140] A 2005 report in the journal *Neurology* recently concluded that estrogen replacement alone increased the risk of Parkinson disease in women who have had a hysterectomy.[141] The statistics are daunting, and have prompted many women to look for alternative methods of relieving menopausal symptoms and protecting the health of their heart and bones.

Alternative therapies

Seventy-five percent of women experience vasomotor symptoms, which are related to the nerves and muscles that cause the blood vessels to constrict or dilate. Common vasomotor symptoms include hot flashes and night sweats. Relief of these symptoms can come in forms that we are not as familiar with in North America, including acupuncture. Several studies show that acupuncture can provide non-hormonal relief of hot flashes and sleep disturbances.[142] Combined with a low calorie diet, acupuncture was more effective at promoting menopausal weight loss and lowering BMI than was a low calorie diet alone.[143]

Phytoestrogens

Phytoestrogens, as we discussed in the hormone chapter, are plant estrogens that are strikingly similar in chemical structure to the estrogen called estradiol. These phytoestrogens bind to estrogen receptors, and exert tissue specific effects. We also know that dietary phytoestrogens are metabolized by bacteria in the intestines, conjugated in the liver, circulated through the blood stream and then safely excreted in urine, likely making them safer than synthetic versions of the hormone because the body knows how to metabolize them.

Many phytoestrogens have been used traditionally for centuries for the relief of menopausal symptoms. Black cohosh, though not currently considered a true phytoestrogen, is used to relieve hot flashes, perspiration, sleep disturbances and depressive mood.[144] Red clover is believed to help in treating several conditions such as hot flashes, and is documented to be useful for cardiovascular health[145]. However, research published in 2004 from the University of Iowa indicated that dietary supplements containing red clover extracts were no better than placebo in regards to hot flashes in menopause.[146] Vitex (chastetree) has also been used to control hot flashes and supports secretion of hormones to balance estrogen.[147]

Take advantage, of other nutrients that will not only relieve your menopausal symptoms, but also promote whole-body health. Anti-inflammatory omega-3 fatty acids – especially fish oils – and the omega-6 fatty acid GLA are useful to relieve breast pain and will help your skin to continue to glow after the skin-protective effect of estrogen diminishes. To combat stress and irritability, you can also use supplemental 5-hydroxytryptophan or 5-HTP (the precursor to the mood enhancing neurochemical serotonin), which will also help with muscle aches and promote deeper sleep. If insomnia is a real problem, try using a high-alpha whey isolate and supplemental melatonin – the body's natural sleep hormone. (See Part Two, Condition 21, Sleep)

Don't underestimate the value of a balanced, nutritious diet in easing menopausal symptoms.

Lifestyle Relief

As mentioned earlier, several symptoms associated with menopause might be related to an underlying issue with adrenal or thyroid gland function. Be sure to address any issues of under-functioning body systems. Other options for natural menopause relief include:

- eating organic meat and produce to avoid exposure to xenoestrogenic pesticides, herbicides and fungicides
- focusing the diet on plant foods
- avoiding coffee and other forms of caffeine to support adrenal gland function
- avoiding alcohol as it has been linked to breast cancer
- drink plenty of pure, filtered water to rehydrate skin and mucous membranes
- support your liver by avoiding fast food, processed food and sugary snacks (this will also help your adrenal glands!)
- quit smoking
- reduce stress. Treat yourself to a regular massage, meditation or walk
- exercise to boost mood and relieve menopausal symptoms
- get sufficient sleep. (See Part Two, Condition 21, Sleep)

Isn't it amazing how many of the above recommendations also boost your metabolism! Coincidence? I think not.

Her Health Protection Chart

Nutrient	Suggested dosage
Ultimate Protein Energy Shake™	1 serving twice daily
Ultimate Her Energy™	2-4 capsules 1-2 times/day
Ultimate Anti-Stress™	2-4 capsules prior to sleep
5-HTP	50-100 mg, up to 3 times/day
Calcium (Glycinate and/or Citrate/Malate)	1,000-1,200 mg/day
Magnesium (Glycinate and/or Citrate/Malate)	600-1,000 mg/day
Multi-vitamin/mineral complex	Follow label instructions
Melatonin	500 mcg-3 mg prior to sleep
X2O mineral sachets	1-2 sachets twice daily in water

thirteen

HIS HEALTH: Andropause

Okay, guys, do you remember what it was like to be a teenager? Do you have memories of that lean, muscular physique that never seemed to gain an ounce of fat no matter how much you ate – or drank? Do you now find yourself patting your belly more then you pat your dog? If you do, you could very well be experiencing a loss of testosterone, otherwise known as andropause.

If the word "andropause" sounds familiar, it's probably because it resembles that transitory time in a woman's life called menopause. Andropause is actually called male menopause in many circles, which, when you think about it, does describe the situation a little better, since we are talking about a pause in a man's life, thus men-o-pause. But the word "andropause" first appeared in literature in 1952 and is commonly defined as the natural cessation of the sexual function in older men due to a marked reduction in male hormone levels, primarily testosterone.

Andropause also happens to be one of the primary reasons some men age prematurely by losing a great deal of muscle while gaining excess fat. Unlike menopause, during which women lose most of their female hormones almost overnight, andropause works ever so slowly, creeping up over many years until you wake up one morning as a completely different person. The man you once were has been replaced by a softer, mushier and sometimes grumpier one.

Testosterone belongs to a group of hormones referred to as the androgens. Testosterone is anabolic and, in men, is responsible for the development of the sex organs and male characteristics, including facial hair, deepening of the voice and muscle development. It also causes aggressive behaviour, plays a role in sperm production and is a factor in libido (remember what that was like in your youth, as well?). Women use testosterone, too, but in different ways. We lose levels of this fundamental hormone as we age, gain weight, become sedentary or neglect our diet – in other words, when our metabolism slows down. One of the first signs of decreasing testosterone is the loss of muscle mass and strength along with a gain in excess body fat, especially in the abdominal region.[148] Abdominal fat also happens to be the most dangerous place to store fat due to its proximity to your vital organs.

Loss of muscle mass is directly related to a decrease in metabolism and the subsequent gain in body fat. In fact, studies show that there is a direct relationship between testosterone and obesity. As testosterone levels increase, lean body mass increases and obesity decreases and vice versa. A 1998 study of 284 middle-aged men, presented in the *Journal Metabolism*, showed that low testosterone levels were both directly and indirectly related to the amount of fat men carried around their midsections.[149] Along with thickening of the arteries, accelerated osteoporosis and decreased muscle strength, a study of 53 men over the age of 50 shows that consequences of andropause include decreased libido, lack of energy, erection problems, falling asleep after dinner, memory impairment, loss of pubic hair, sad or grumpy mood changes, decrease in endurance, loss of axillary hair and deterioration in work performance.[150,151]

All Tied Up

You certainly can't ignore the research when it comes to the antiaging, fat-fighting potential of testosterone, but not all testosterone is created equal, because not all testosterone in the body is biologically active. It is only the free, unbound testosterone that exerts its wondrous effects on men's bodies. It is also the free, physiologically active testosterone that declines the most with age.

The decline in free testosterone is due to the increased binding effect with age of a transport protein called sex hormone binding globulin (SHBG).[152] Testosterone is unable to elicit its physiological responses once it is bound to SHBG. It is not uncommon for free testosterone to decline by about 1% per year after the age of 35, and it is interesting to note that SHBG is regulated in part by insulin levels. This implies that dietary carbohydrate levels, blood sugar and blood insulin actually affects testosterone levels over the long haul. This is another reason to be careful about too many carbohydrates.

One natural herb that will assist in this goal is stinging nettle. In a paper published in 1995 in the journal *Planta Medica*, a proprietary extract of stinging nettle plant (found in *Ultimate Male Energy*™) was shown to inhibit testosterone binding to SHBG.[153,154] According to this and other studies, stinging nettle extract is not only a natural booster of testosterone, but can also help to prevent and perhaps treat prostate disease.[155]

The Balancing Act

Losing testosterone or having it bound up are far from the only problems men face as they get older. As men experience a drop in their testosterone levels, or a rise in their SHBG levels, they usually experience a rise in their estrogen levels, too.[156] Estrogen is the anabolic hormone responsible for development and maintenance of female sex characteristics and play a role in fertility, reproduction, menstruation and menopause. Estrogens are also a factor in the growth of long bones. (Men create and use estrogen as well, but not nearly as much as women.) It is well established that tissue accumulation of estrogen is a unique hallmark of andropause.[157]

It turns out that as men lose muscle and gain body fat, their fat cells manufacture a special enzyme called aromatase. Aromatase is responsible for converting valuable testosterone into estrogen and the more fat we accumulate, the more estrogen is produced. In fact, it is not uncommon for a man of retirement age to have higher estrogen levels in his body than a woman of the same age![158] (Talk about getting in touch with your feminine side). And that's not a good thing. Higher estrogen levels are associated with prostate problems as well.[159] (See Part Two, Condition 19, Prostate Health.) The good news is that a natural proven aromatase inhibitor exists and according to a study appearing in the *Journal of Steroid Biochemistry*

& Molecular Biology, the flavonoid chrysin is similar in both potency and effectiveness to a pharmaceutical aromatase inhibitor.[160] (Chrysin is just one of the ingredients in *Ultimate Male Energy*™)

Estrogen from all directions

Estrogen dominance relates to hormonal imbalances that tip too much in the direction of estrogen in one's body. Estrogen dominance isn't just a result of a decline in androgens, but is also a consequence of dietary factors and lifestyle choices. Xenoestrogens (false estrogens) in the environment share some of the responsibility. (See Chapter 3, The Hormone Connection.) Sources of xeno-estrogens include commercially raised (non-organic) beef, chicken and pork, pesticides, herbicides, canned foods and lacquers. Other sources include spermacide, detergents, plastic drinking bottles and personal care products like shampoo, conditioner and shaving cream. Alcoholic beverages are also known to raise circulating levels of estrogen by as much as 200%, and even one night of drinking with the boys can lower testosterone for 24 hours.[161]

What's a guy to do?

To reduce the impact of estrogen in your body all the while maxi-mizing testosterone levels:

- use glass rather than plastic to store food
- never heat food in plastic containers in the microwave (if you must use one)
- buy organic produce and meats
- use natural pest control around your home rather than pesticides
- avoid synthetic chemicals in your personal care products
- use condoms without spermicide
- minimize intake of soy, a known plant estrogen found in the majority of processed foods on store shelves
- consume sufficient quantities of high quality protein
- perform regular bouts of resistance training (weight training)
- get sufficient sleep
- Take testosterone supporting and negative estrogen producing nutrients (found within *Ultimate Male Energy*™)

Keep your system clean

Remember that estrogens are metabolized in the liver, so it is imperative to support your liver before the estrogens are sent to the intestines to be eliminated. (See Part Two, Condition 11, Liver Health.) Avoid alcohol and other toxins, including cigarettes. Studies show that smokers reach andropause earlier than non-smokers. Eat a metabolism-boosting diet as described in Chapter 8. Milk thistle and dandelion root are known liver tonics. Include plenty of fibre on the menu, as fibre helps to pull estrogen through the digestive tract for disposal from the body. A stagnant bowel gives estrogen time to be reabsorbed into the bloodstream.

And give testosterone a boost

Natural testosterone-boosting nutrients help to block aromatase (which you will recall is responsible for converting testosterone into estrogen), free testosterone or to block estrogen metabolites. Include nutrients like chrysin, stinging nettle root and indole-3-carbonol (all found within the formula *Ultimate Male Energy™*), plus zinc in your daily supplement regimen to keep estrogen in its place. And you surely remember from your earlier reading that weight training helps to boost testosterone levels. Pump it up!

His Health Protection Nutrient Chart

Nutrient	Suggested dosage
Ultimate Protein Energy Shake™	1 serving twice daily
Ultimate Male Energy™	2-6 capsules/day
Multi-vitamin/mineral complex	Follow label instructions
Zinc (Picolinate, OptiZinc® or citrate)	12-25 mg/day
X2O mineral sachets	1-2 sachets twice daily in water

fourteen

LIBIDO

Men hit their *so-called* peak in their late 20's, however this peak is most often defined by a man's energy, stamina and of course his sexuality! The truth is that a man's (as well as a woman's) sexuality is largely controlled through the production of the powerful hormone testosterone, (which is in abundance in a man's body between the ages of 18 and 30).

As you learned in the last chapter (His Health) one of the consequences of a drastic reduction in testosterone levels (also referred to as Andropause) is a greatly diminished libido. Studies indicate that men can lose approximately 10 percent of their testosterone levels each decade after age 30.

Aside from the fact that overall health is negatively affected as we lose this vital hormone, this dip in testosterone also leads to low mood, loss of muscle mass and strength, and a drastic decline in the libido that once never failed us. In other words, testosterone is the primary hormonal message behind a man (and a woman's) *sex drive.*

The grim fact remains that at least 52 percent of men past the age of 40 suffer from some degree of erectile dysfunction or ED (the inability to achieve and/or keep an erection sufficient for intercourse). The problem is that when you take away a man's ability to both achieve and/or maintain a powerful erection, you also strip that man of his confidence, which often leads to feelings of inadequacy, frustration and a lowered love for life. In other words, ED can often be life threatening to men.

Prostate disease, your heart and ED

Studies also indicate that well over half of all men who suffer from prostate disease also suffer from erectile dysfunction and other sexual problems. This makes a lot of sense once you understand that the prostate is an integral part of a man's sexuality and proper ejaculation is dependant upon the health of the gland.

There is even speculation that ED may be an early sign of heart disease. Researchers from the University of Minnesota School of Medicine believe that anyone experiencing ED should be immediately assessed for cardiovascular risk. Let's face it – the body isn't going to concentrate fat deposits in only one area of the body!

It is by no mere coincidence that *erection-enhancing* medications like Viagra® are experiencing sales through the stratosphere! And it is also no surprise that men who use these drugs often experience a new zest for life (not to mention their partner's new found happiness). The problem is – with any substance not part of our evolutionary biochemistry – the majority of these erection-supporting drugs can also carry with them numerous side-effects. Figures show that one out of every five men over the age of 40 has tried Viagra with 48 percent experiencing at least one side effect. Aside from this, research from the School of Osteopathic Medicine in Stratford, USA, indicates that one-third of patients who use erectile dysfunction drugs report unsatisfactory results.

Getting it done with testosterone

The majority of ED drugs today concentrate on delivering enough blood to the penis to experience an erection, but they do very little to nothing to increase sexual arousal. Just because the blood is pumping, doesn't mean you're in the mood! This is where testosterone comes in. As mentioned, testosterone is the key hormone of desire. Studies show that low testosterone levels can often lead to ED, which is especially true in diabetics. Many researchers have shown that testosterone therapy can be a viable option in the treatment of ED. One study appearing in the *International Journal of Impotence Research* indicated that testosterone therapy was very effective in those who were unable to experience results with Viagra, and was even more effective when combined with Viagra.

Raising testosterone with a little help from Mother Nature

Aside from following a healthier diet and exercising more – especially resistance exercise (which has been shown to help elevate testosterone levels) – nature has provided a few natural substances that can often give low testosterone levels, and thereby low libidos, a little extra lift (no pun intended). I'm not talking about the numerous all *hype* with little to no *action* libido enhancing supplements that crowd store shelves these days, but instead proven nutrients – used by various cultures around the world for centuries – that truly provide results!

One such nutrient is an herb grown in Southeast Asia called Tongkat Ali, which is believed to be one of the most powerful aphrodisiac plants on earth. Researchers from the School of Pharmaceutical Sciences, in Malaysia, discovered that the root of Tongkat Ali is a potent stimulator of sexual arousal and the Asian Congress of Sexology published a paper in 2002 touting the incredible aphrodisiac and testosterone boosting powers of this amazing herb.

Aside from this, the *British Journal of Sports Medicine* reported in 2003 that Tongkat Ali was able to increase muscle mass and reduce body fat in men possibly by increasing testosterone. Needless to say, this incredible herb has gotten many a man, *"excited!"* Further research using Tongkat Ali that was presented in the Partial Androgen Deficiency of the Aging Male (PADDAM) study in 2005, using thirty men between 31-52 years of age, showed an increase of testosterone in all subjects along with a 91 percent improvement in libido and a 73 percent improvement in sexual function.

Ultimate Libido is a synergistic formula that incorporates *effective dosages* of *properly extracted* nutrients like Tongkat Ali to help boost libido naturally and side-effect free.

It all starts in your head

Sexual arousal is not solely allocated to the genitals; it is actually a whole body experience that starts in your brain. In both man and woman, the sexual message is transmitted via a brain chemical or neurotransmitter called acetyl-choline (ACH). When ACH levels are low we often experience low libido and visa versa.

The good news is that we can fairly easily elevate bodily levels of ACH by consuming supplements of choline – especially in bioavailable forms like phosphatidylcholine (otherwise known as lecithin), Alpha-glycerylphosphoryl-choline (alpha-GPC) and CDP-Choline (cytidine diphosphate choline). ACH levels can also be elevated by consuming extra vitamin B5.

Libido Protection Nutrient Chart

Nutrient	Suggested dosage
Ultimate Libido™ (with Tongkat Ali)	3 capsules in the morning for 1 month and another 3 capsules approximately 1 hour prior to sexual activity
Choline	1000-3000 mg
Vitamin B5 (pantothenic acid)	100-500 mg

fifteen

LIVER HEALTH

When it comes to the integrity of your cellular metabolism, one of the most important areas of concern lies within a very powerful, yet often overlooked organ – the liver. Not only is your liver the largest of your internal organs and responsible for performing over

500 different chemical functions, it may also be your body's chief metabolic organ responsible for burning great amounts of fat!

Located under the diaphragm on the upper right side of the abdomen, the liver weighs approximately three pounds and is "power central" in your body. Although we are still learning what contributions the liver makes, some of its responsibilities are:

- converting our food into stored energy (glycogen) to fuel muscles
- converting food into chemicals necessary for anabolism and catabolism
- processing drugs, medications and poisons (including alcohol) to prepare them for disposal
- regulating blood sugar levels
- breaking down ammonia (created by the metabolism of certain proteins) and creating urea, which is excreted by the kidneys
- creating bile (stored in gall bladder) to metabolize fats and allow absorption of fat-soluble vitamins A, D, E and K
- manufacturing blood, clotting factors, proteins and in excess of a 1,000 different enzymes.
- hormone metabolism and regulation
- converting thyroid hormone T4 to metabolically-active T3
- manufacturing approximately 50% of the body's cholesterol

And this is just the short list.[162] The health of your liver is serious business. In fact, without it, we would die within 24 hours.

Numbers show that liver disease becomes more common as we age, and if you've been following our discussion of metabolism, that makes sense. As a sedentary lifestyle and inadequate nutrient intake takes its toll, declining metabolic function shows itself in the liver with a decrease in liver volume, declines in drug metabolism, shifts in the synthesis of various proteins, diminished rates of DNA repair and telomere shortening. (See Chapter 3, The Hormone Connection.) As a result, as we age, the liver becomes more catabolic and is thereby less able to rejuvenate, so we experience an increased susceptibility to liver disease.[163] Symptoms of liver trouble include yellow discoloration of the skin or eyes (jaundice), prolonged bouts of itchy skin, dark urine or stool that is pale or bloody and tar-like, severe abdominal pain and swelling, chronic fatigue, nausea, loss of appetite and weight gain.

When it goes wrong

When normal liver cells become damaged and replaced by scar tissue, the result is a condition called cirrhosis. In cirrhosis, the scar tissue distorts the normal liver structure and interferes with the flow of blood, leading to impairment of liver function. Conditions that lead to cirrhosis include chronic viral hepatitis (B, C and D), an autoimmune liver disease called primary biliary cirrhosis, inherited congenital disease, a disease of the bile ducts called sclerosing cholangitis, and the most common cause, excessive alcohol intake. Other causes include heredity accumulation of excess copper (Wilson's disease) and iron overload (hemochromatosis.) People with cirrhosis often notice loss of appetite, vomiting and weight loss. Swelling of the back, legs and feet, as well as a fluid buildup in the abdomen, are also common. Managing cirrhosis requires a protein balancing act: the liver requires adequate protein to promote healing, but too much will lead to excess ammonia in the blood, which will damage the liver further. Because sodium can lead to fluid retention, a low sodium diet (minus the processed foods but high in fresh fruits and vegetables) is recommended.

Hepatitis A

Hepatitis A involves inflammation of the liver caused by the hepatitis A virus (HAV). HAV is typically transmitted by food or drink that has been contaminated with the stool of a person with HAV, and is more easily spread in areas with poor sanitary conditions. It can also be transmitted sexually from men to men.

Hepatitis B

Once thought to be transferred only through blood products, hepatitis B virus (HBV) can also be transmitted through body piercing and tattoos using un-sanitary instruments. Sexual contact and childbirth can also lead to HBV infection.

Hepatitis C

Inflammation of the liver due to hepatitis C virus (HCV) is the most transfusion associated hepatitis and can lead to cirrhosis and cancer. Transmission by sexual contact is rare. An estimated four million Americans have been infected with HCV, and about 85% will develop chronic liver disease. Between 8,000 and 12,000 people will die this year as a result of the disease.[164,165]

Fatty Liver Disease

Despite what its name suggests, fatty liver disease is not caused by eating too many fats. Instead, this condition refers to excess fat inside the liver cells. Nutritional causes of this condition, include protein deficiency and starvation. It is also linked to intestinal bypass surgery for obesity, hormone disorders and obesity. In turn, fatty liver is also associated with free radical damage, mitochondrial dysfunction, inflammation and other factors leading to damage of the liver cells and progressive liver disease.[166] Non-alcoholic fatty liver disease (NAFLD) is associated with obesity, and the best treatment is gradual weight-loss using a well-balanced, nutrient-rich diet and avoiding alcohol intake.

Live long with your liver

Do you ever have one of those days when you are sure to miss a work deadline, the kids need to be at two different places on opposite sides of town at the same time, you forgot to eat so you have a whopping headache, and just as you pull your car onto the street to chauffeur your little darlings, the sputtering and gasping of your car lets you know that you also forgot to fill it with gas? As you wait for the auto club to rescue you, do you remember thinking "This is the straw that broke the camel's back?" (or something like that?) Anyway, it's the same with your liver. Think of all the things it has to do every second of every hour of every day (filtering medications, pesticides, herbicides, fertilizer, hormones, artificial sweeteners, MSG, birth control pills, alcohol, etc.) whether you are asleep or awake. The best thing you can do for your liver is to limit the burdens you place on it so this vital organ doesn't collapse under the workload. Avoid over-indulging in alcohol, and limit your intake of xenobiotics by purchasing organic foods as often as possible. (See Chapter 3, The Hormone Connection.) Instead of saturated fats from meats and dairy foods, choose fish oil, olive oil, walnut oil and flax oil for your fatty acid sources.

Address issues with insulin-resistance, as research shows a strong link between metabolic syndrome and cirrhosis of the liver. Research performed in Japan in 2005 showed that 29% of those suffering from non-alcoholic fatty liver disease also had impaired glucose metabolism.[167] (See Part Two, Condition 17, Metabolic Syndrome.) A high fibre diet is essential for liver health, so be sure to increase your intake of high fiber fresh fruits, vegetables and sprouted whole grains. Legumes like kidney beans and peas contain the amino acid arginine, which helps to detoxify ammonia. Fresh vegetable juices containing apple, carrot and beet help to cleanse the liver (but be careful with these if you suffer from excessive weight gain or are insulin resistant). Artichoke, dandelion root, milk thistle and D-glucarate can also be very beneficial for liver health. And don't forget water. Water primes the metabolic pump, keeping all systems running smoothly.

While I've got you thinking about running, research shows that exercise helps to lessen glucose intolerance in the liver associated with fatty liver disease.[168] It makes sense, then, that exercise would help to prevent it, too! Lace up your sneakers and head outside for 30 minutes of exercise a few days per week. And long live your liver.

Liver Protection Nutrient Chart

Nutrient	Suggested dosage
Multi-vitamin/mineral complex	Follow label instructions
B vitamins (high-potency)	Follow label instructions
Vitamin C (as mixed ascorbates)	500-2,000 mg/day
Vitamin E (as mixed tocopherols)	800 IU daily
Probiotics	Follow label instructions

sixteen

LUNG HEALTH: Asthma and Bronchitis

As mentioned throughout this book, oxygen is the most important substance for the human body. Oxygen is also what drives our metabolism, right down to the cellular level. But for many, the act of breathing is not as easy as it should be. We often go about our days inhaling and exhaling without much thought, until a stuffy nose, bout of bronchitis, sinusitis or an asthma attack takes our breath away. According to the World Health Organization, asthma is a serious health threat affecting over 100 million people worldwide. In Canada, 8.4% of the population has been diagnosed with asthma, and the number of asthmatics in the US has increased by over 60% since the early 1980s. The global death toll of this condition exceeds 180,000 every year.[169]

Although conditions like asthma, bronchitis and sinusitis seem like respiratory problems because they affect breathing, it is probably better to think of them as inflammatory conditions. Any condition that ends in "itis" is an infection, and infection always involves inflammation. (Asthma doesn't end in "itis" and is not an infection, but nonetheless also involves inflammation.)

Asthma is often an allergic reaction to an inhaled particle or food that sets off a chain of events leading to inflammation. Throughout the body, we have specialized immune cells called mast cells, which are covered with Immunoglobulin E (IgE) antibodies. When an asthma trigger finds its way to your mast cell and the IgE antibodies, the mast cell fires its contents into the surrounding tissue. As a result, a host of inflammatory messengers called leukotrienes are released, including histamine and other potent biochemicals.

In an asthmatic episode, the muscle tissue in the walls of the bronchi start to spasm, causing the cells that line the airways to become inflamed and secrete mucus into the air spaces. Together, these actions cause the bronchi (the large air tubes that convey air to and from the lungs) to narrow and breathing becomes much more difficult. The airways of asthmatics become hyper-reactive to stimuli that don't affect healthy lungs.

Repeated asthma attacks promote an abnormal thickening and hardening of air passages and the immune system secretes immune factors that eventually destroy delicate tissues lining the airways.

Bronchitis involves inflammation of those tubes that transport air to the lungs, and can be acute or chronic. Inflammation stimulates the production of mucous (sputum) that can add to the obstruction of the airways and increase the likelihood of bacteria infections in the lungs. Chronic bronchitis entails a daily cough that produces sputum for three months in two consecutive years. Long-term consequences of bronchitis include an increased risk of cardiovascular disease, showing how important oxygen is to your heart and blood vessels. Damage to heart function appears to be amplified in those with obstructive breathing disorders who also have an existing underlying cardiac rhythm problem.[170] This makes sense when you consider that our bodies need oxygen to achieve optimum metabolism and breathing disorders obviously reduce the availability of oxygen in the bloodstream.

In sinusitis, the membrane lining the hollow areas in the bone of the skull near the nose (the nasal pharynx) becomes inflamed. Known as sinuses, these hollow areas are connected to the nasal cavities. Thickened mucous caused by dehydration, disease, antihistamines or lack of humidity in the air can block drainage of the cavity leading to stagnation. Stagnated mucous is a ripe breeding ground for bacteria and fungus. Bacterial pathogens involved in acute sinusitis include *Streptococcus pneumonaie, Haemophilus influenza* and *Moraxella cararrhais.* Staphylococcus aureus is predominant in chronic sinusitis. Sinusitis is the most common chronic illness in the United States.[171,172,173]

Take a deep breath

Helping to control inflammation in the body is a function of what we eat and the amount of exercise we get. Eating fresh fruits and vegetables keep our trillions of cells strong so they can fight off irritants and disease. Vitamins and herbs like the curcuminoids from standardized turmeric extract prevent and squelch damaging free radicals. Turmeric has been traditionally used for centuries to prevent the release of histamine and fight inflammation in asthma. Research shows that a soy isoflavone called genestein is associated with better lung function in asthma patients, while high intake of margarine can increase the risk of clinical onset of asthma in adults.[174,175] A review of studies concerning asthma and obesity (which often occur together) suggest that consuming anti-inflammatory EFA gamma linolenic acid from borage oil and eicosapentaenoic acid from fish are beneficial for managing asthma.[176] In fact, some research suggests that eating fish in childhood can prevent the onset of asthma.[177] Again, this is hardly surprising since we already know that these oils are important to cellular metabolism.

Researchers from the Australian Commonwealth Scientific and Industrial Research Organization (CSRIO) were not only able to drastically extend the life of both lung and skin cells by adding the dipeptide *carnosine* to the cell medium, but also caused cells to remain in a youthful state right up until senescence (cell death). When these cells were replaced in a regular cell medium, they instantly started to show signs of aging. Consider supplementing with carnosine for the life of your lungs. And don't underestimate the power of simple vitamin C. With properties as an anti-inflammatory nutrient as well as an antioxidant, vitamin C is an excellent partner in your healthy breathing.

Bottom line, the same foods that promote a healthy metabolism also protect respiratory health. Focus on fruits, vegetables, lean protein and sprouted whole grain foods while avoiding alcohol and fatty foods. And be sure to get enough to drink. Dehydration is a factor in asthma, sinusitis and myriad other health complaints.[178] Opt for a minimum of eight eight-ounce glasses of filtered water daily.

Moving your metabolism

Exercising regularly also keeps your lungs healthy. Although exercise is sometimes a trigger for asthma and other breathing conditions, the overwhelming majority of studies demonstrate the value of those with breathing disorders participating in moderate exercise. Benefits in cardiovascular health and overall quality of life are considered of prime importance.[179] Avoid exercising outdoors during extreme weather conditions.

Lifestyle concerns

Although it should go without saying, avoiding asthma attack triggers is a must to break the inflammatory cycle. Use an air filter in your home, and for goodness sake, stay away from smokers (because I know you don't smoke yourself!) If you suspect that food might be a trigger for your attacks and are having trouble isolating it, consider working with an allergist or alternative care practitioner. Avoid animal fat, especially beef, pork and organ meats.

"A full 96% of sinus infections are fungal, and 15% of those infections were identified as *Candida*. It is possible to reestablish control of *Candida* through dietary changes, including the elimination of all forms of sugar for a period of time."

If sinus infections are the bane of your existence, it is possible that you are dealing with a systemic *Candida* infection. *Candida* overgrowth results from an imbalance in the intestinal flora and can migrate out of the digestive tract to wreak havoc in tissues throughout the body. A full 96% of sinus infections are fungal, and 15% of those infections were identified as *Candida*. It is possible to reestablish control of *Candida* through dietary changes, including the elimination of all forms of sugar for a period of time. (See Part Two, Condition 10, Gut Health.)

Lung Protection Nutrient Chart

Nutrient	Suggested dosage
Ultimate Protein Energy Shake™	1 serving twice daily
Ultimate Longevity (with carnosine)™	1 capsule/day
Ultimate Fatty Acid Complex™	4 capsules 2-3 times/day
Multi-vitamin/mineral complex	Follow label instructions
B vitamins (high-potency)	Follow label instructions
Vitamin C (as mixed ascorbates) with quercetin	500-2,000 mg/day
Vitamin E (as mixed tocopherols)	800 IU daily
Probiotics	Follow label instructions

seventeen

ORAL HEALTH: Gum disorders

For another brilliant example of how the health of one part of your body impacts every one of your trillions of cells, you have to consider the health of your mouth. While it's obvious that the mouth is important for metabolism simply because it is the entry point for our food, the mouth also has the potential to be the site of origin for significant health risks with far-reaching consequences. In other words, put your metabolism where your mouth is!

Cavities are only the beginning

The first sign of dental trouble is often a cavity. Cavities are the result of many factors, including plaque (consisting of microorganisms, skin cells and immune system cells) that can harden into tartar or calculus if not removed regularly. Bacteria ferment dietary sugars found in plaque (remember anaerobic metabolism from Part Two, Condition 6, Cancer), creating an acid that is harmful to teeth and can lead to a cavity. If a cavity is left untreated, the tooth will continue to decay until an infection sets in.

Plaque can also trigger inflammation of the gingiva, more commonly known as the gum. When gingivitis occurs, affected gums become painful and swollen, and bleeding is common. Inflamed gums have also been linked to a deficiency in vitamin C and co-enzyme Q-10 (co-Q10). If left untreated, gingivitis can progress to periodontitis, occurring when invading bacteria spread to the bone underlying the gum. Periodontitis always requires medical attention. Often very painful, this condition can lead not only to the complete loss of the tooth but it is also linked to diabetes, bacterial pneumonia, respiratory conditions and serious heart infections.[180,181,182]

Brushing for your Heart

Endocarditis is an infection of the endocardium, which is the lining of the interior surface of the heart chambers. (The endocardium is made up of a layer of endothelial cells covering a layer of connective tissue.) This infection occurs when bacteria in the bloodstream find their way to the heart. When gums are damaged by illness and injury, bacterial passage into the bloodstream is simplified. Bacteria accumulate into "clots" which can travel throughout the bloodstream, increasing the risk of heart attack and stroke. While people with damaged heart valves are most often affected, normal heart valves are also at risk for endocarditis.[183] As a preventive measure for endocarditis, patients with certain heart conditions are given preventive antibiotics at regular dental visits. The bottom line is, for the sake of your heart's metabolic function, you need to pay attention to your mouth.

Let's talk about brushing

I know; you've been brushing your teeth by yourself since you were two years old. What can I possibly have to tell you about brushing that you don't already know? Actually, I'd like to talk about what you've been brushing your teeth *with*. We've heard our whole lives (since we were two!) that we need a fluoride toothpaste to prevent cavities. You might be surprised to learn that theory has come under fire.

In many communities throughout North America, fluoride has been added to the water supply to help prevent cavities. The problem, however, is that fluorine is a poisonous and corrosive trace element. It has been linked to bone disease including fluorosis, cancer and osteoporosis, reduced brain activity and damage, gastrointestinal problems, and depressed thyroid function to name but a few concerns. It also causes dental fluorosis, the mottling and pitting of developing teeth. (Don't forget that cavities form in pits and crevices.) While some might call this a cosmetic concern, damage can be so severe that teeth crumble and fall out. (So what does it do to your insides?) Current estimates predict that 48% of children living in fluoridated areas now have some form of dental fluorosis.[184]

And we don't need it in the water. Recent studies throughout the world show the rate of cavity-development does not increase in localities that remove fluoride from the water-supply, probably because we are taking better care of our teeth.[184] But here is another problem: as usual, our society always thinks more is better. If fluoride, theoretically, is good for teeth, we'll put it in everything. You can now find this corrosive chemical in toothpaste, mouthwash, chewing gum and, because it is also in the water supply in many agriculture communities, fluoride is also present in our fruits and vegetables. Because it is a natural element in the earth, we already get it in our produce through the soil. Enough already! For the sake of your health, opt out of the fluoride craze. Visit your local health store and find a fluoride-free toothpaste for you and your family. Along with regular flossing, brushing your teeth without fluoride will keep them healthy and strong.

Food for thought

It should come as no surprise that the health of your mouth depends a lot on what you put into it. Research from the School of Dentistry and Biochemistry at the University of Maryland shows that deficiencies in antioxidants coupled with insufficient dietary protein promotes impaired immunity and a reduction in the efficiency of our salivary glands. Saliva is essential not only for digestion, but also as a means of keeping the mouth clean. In fact, a lack of saliva is a major cause of tooth decay. A deficiency in antioxidant vitamins and minerals is also linked to the development of oral cancers.[185] Adequate vitamin A is essential for the health of your gums and lining of your mouth. Vitamin C helps to prevent bacterial growth, and your teeth benefit from sufficient calcium and vitamin D.

Coenzyme Q 10 is a fat-soluble nutrient that works as a messenger link between various enzymes in the mitochondria of the cell, making it invaluable in energy metabolism. CoQ10 is also a significant antioxidant that can neutralize free radicals that threaten the cell membrane. Research shows that tissue levels of CoQ10 are deficient in patients with peridontitis, and that gum health of heart patients improved with supplemental CoQ10.[186] Since no other substance can substitute for CoQ10, it's a good idea to supplement with it daily. And perhaps you can rinse it down with green tea. Green tea contains two very powerful polyphenolic compounds called catechins that have been shown to combat dental plaque and bacteria. These catechins, called epigallocatechin gallate (EGCG) and epicatechin gallate (ECG), can also restrain the activity of collagenase. Collagenase is a naturally-present enzyme that goes out of control when it is around bacteria, and ends up destroying healthy collagen in the gums. EGCG and ECG also inhibit *Streptoccus mutans* bacteria, one of the most common bacteria for causing cavities. In fact, since you aren't using fluoride toothpaste anymore, why don't you consider brushing with green tea? In a Chinese study, green tea extract was used to brush and rinse the teeth. Only five minutes of contact with the green tea completely de-activated *Streptoccus mutans.* Researchers concluded that green tea helps to prevent cavities.[187] Remember that when you are at a restaurant with friends. Be sure to order "a cuppa" green tea to finish your meal!

Make sure your metabolism-boosting diet focuses on fresh fruits and vegetables, lean sources of protein, whole (preferably sprouted) grains, nuts and seeds. Prevent cavities through proper oral care, and be sure to brush at least twice a day and floss daily.

Oral Health Protection Chart

Nutrient	Suggested dosage
Ultimate Protein Energy Shake™	1 serving twice daily
Ultimate Longevity (with carnosine)™	1 capsule/day
Multi-vitamin/mineral complex	Follow label instructions
Co-Q10	30-100 mg/day
Vitamin C (as mixed ascorbates)	500-2,000 mg/day
Vitamin E (as mixed tocopherols)	800 IU daily
Multi-vitamin/mineral complex	Follow label instructions
B vitamins (high-potency)	Follow label instructions
Calcium (Glycinate and/or Citrate/Malate)	1,000-1,200 mg/day
Magnesium (Glycinate and/or Citrate/Malate)	600-1,000 mg/day
Vitamin D3 (Cholecalciferol)	800 IU daily
Vitamin K2	15 mg, 1-3 times/day
Probiotics	Follow label instructions

eighteen

METABOLIC SYNDROME

Let's take another sentimental journey into your childhood, when life was full of energy and your healthy cells were radiating vitality. Now fast forward a few years in your memory. Do you recall that day when you woke up and realized that suddenly your body didn't perform as well as it did the day before? I mean one day you sprung out of bed at the crack of dawn and moved non-stop from morning to night, and the next day you preferred to hang out on the couch, drinking coffee and lamenting your lack of time for exercise? Oh, it didn't happen to you overnight? Perhaps you remember instead that you experienced a gradual change in energy as the years passed? The important thing to remember here is that we don't wake up one morning with a serious illness (our bodies don't just decide to malfunction!). Instead, it often takes years for cellular metabolism to erode to the point that disease takes over.

On the way to disorders like diabetes and cardiovascular disease (CVD), your metabolic systems gradually start to change. To make us aware of these changes, the body starts to provide clues that all is not right inside. Metabolic syndrome (MetS), a new epidemic on the health care horizon, is one of those clues. Although you might not have heard of it yet, approximately 47 million Americans suffer from MetS. Most frightening, an estimated one million young people from the ages of 12-19 are currently living with the condition.[188]

"Although you might not have heard of it yet, approximately 47 million Americans suffer from MetS."

Also known as dysmetabolic syndrome, the insulin resistance syndrome and syndrome X, MetS refers to the clustering of risk factors for diabetes and CVD, including impaired insulin sensitivity, hyperglycaemia, elevated blood fats, abdominal obesity and high blood pressure.[189,190] Insulin resistance, you will recall from the chapter on diabetes, means that the cells in your muscle, fat and liver do not use insulin properly. Glucose levels rise in the blood-stream, leading to serious health consequences, including kidney failure, blindness and amputations. I wrote about the effects of obesity and high blood pressure on health in the heart disease (Part Two, Condition 11, Heart Health.) These are not conditions that you want to casually walk into. Risk factors for MetS include family history of diabetes or coronary artery disease in first or second degree relatives, gestational diabetes (only occurs during pregnancy) or giving birth to a baby weighing nine pounds or more – ouch![191]

Diagnosing metabolic syndrome:

According to the National Cholesterol Program[192], the presence of three or more of the following conditions indicates syndrome X (MetS):

- excess weight around the waist (more than 35 inches for women and more than 40 inches for men)
- low levels of HDL (good) cholesterol (below 40mg/dL for men and below 50mg/dL for women)
- high levels of triglycerides (150mg/dL or higher)
- high blood pressure (130/85 mm Hg or higher)
- high fasting blood glucose levels (110 mg/dL or higher)

Blood tests determine cholesterol and triglyceride levels. A glucose tolerance test lets you know about your blood glucose levels. It works like this: you drink a sugar solution to determine how your cells metabolize the glucose. Blood levels are measured two hours after you have your drink. Current guidelines indicate that if your glucose levels exceed 126 mg/dL on two consecutive occasions, you are probably diabetic, whereas levels over 109 mg/dL indicate pre-diabetes or insulin resistance. Optimal glucose ranges should be below 100 mg/dL.[193] The fact remains that a healthy pancreas will continue to produce insulin until blood glucose levels drop below 83 mg/dL.[194] According to the Life Extension Foundation in Florida, optimal blood glucose levels should not exceed 86 mg/dL.[195]

Putting metabolic syndrome in reverse

If you suspect or know that MetS is a factor in your life, you are not on a one-way street to diabetes. In fact, there is a lot you can do to keep diabetes and heart disease out of your future. All you have to do is make some lifestyle changes. As you've probably guessed, you need to eat a nutritious metabolism-healing diet and add some exercise back into your life.

A major clinical trial called the Diabetes Prevention Program was performed by the National Institutes of Health. Researchers chose 3,234 overweight participants who had impaired glucose tolerance (two markers of MetS.) Participants were divided into four groups: one whose treatment focused on diet, lifestyle and behaviour modification; two that took medications; and one group that took a placebo medication. The lifestyle group aimed to lose seven percent of their body weight and maintain that loss by eating less fat and fewer calories as well as exercising for 150 minutes per week. During the trial, one of the drugs proved to be toxic to the liver and its use in the program was halted. Participants in the lifestyle intervention group – those receiving intensive counseling on effective diet, exercise and behavior modification – reduced their risk of developing diabetes by 58 percent. The benefits of lifestyle modification were so quickly evident that the study was actually halted a year early. Researchers concluded that weight loss through diet and exercise reduces the risk of diabetes by improving the body's ability to use insulin and process glucose – and that the lifestyle factors were more effective than drugs! This is an important piece of information, considering that people who have MetS are likely to develop type 2 diabetes within 10 years, unless they take action to prevent it. Other studies show that MetS increases risk of CVD by 1.5-6 fold.[196,197]

Taking the prevention advice a bit further, if you are currently at a healthy weight, you can also prevent the development of MetS through diet and exercise! Aim for 30 minutes of exercise like walking or bicycling five days a week, coupled with resistance weight training at least twice weekly, and eat according to the principles outlined in Chapter Eight. Your metabolism will reward you with years of healthy vitality!

Metabolic Syndrome Prevention Chart

Nutrient	Suggested dosage
Ultimate Protein Energy Shake™	1 serving twice daily
FibreLean™	1 serving, 3 times/day with a full glass of water
Ultimate Longevity (with carnosine)™	1 capsule/day
Chromium (as LeafBrand)	400-1,000 mcg/day
Multi-vitamin/mineral complex	Follow label instructions
X2O mineral sachets	1-2 sachets twice daily in water

nineteen

OBESITY: The True Epidemic

According to a 2002 study published in the *Journal of the American Medical Association*, nearly two-thirds of the U.S. population is overweight as measured by the body mass index or BMI (see Defining the Problem).[198] As staggering as this information is, further reports from the U.S. Department of Health and Human Services shows that nearly 300,000 people die each year in the U.S. alone because of unhealthy dietary habits and physical inactivity or sedentary behavior that leads to overweight and obesity.[199] Trust me when I say that Canadians aren't very far behind.

Research from *Statistics Canada* in 2005 shows that in the last eight years alone, 1.1 million Canadians have joined the ranks of the obese. The study indicated that 33% of Canadians of normal weight in 1994/1995 were overweight by 2002/2003. Thirty-three%![200] That's a huge increase – pun intended! In this day and age, it may be normal to be fat, but when did normal ever equate to true health? It's only normal because this is what the majority of our population experiences. Healthy, energetic, fit and lean doesn't just happen effortlessly. It takes dedication and commitment to maintain proper metabolism and live the life you were meant to live.

A Bite at a Time

Do you ever find yourself glaring at someone who lights up a cigarette beside you on a park bench, annoyed that they aren't giving a second thought to the second-hand smoke now blowing in your direction? Now, imagine that someone sits down beside you, and pulls out a fast food burger from a bag. Can you hear yourself ask: "What do you think you are doing eating that around me? It's bad enough you are doing it to yourself, but do you need to subject me to it as well?"

We all know that smoking kills. There is endless research proving it. Disturbing images on cigarette packages in stores across the country are enough of a daily reminder. Now, think of your reaction if the next time you open a box of your favourite donuts, (with all those tantalizing flavours, colours, glazing and sprinkles), there were horrific pictures on the inside flap of the box, showing cholesterol-clogged arteries, insulin syringes and malignant tumors, with a warning label that read: "Eating donuts can cause obesity, diabetes, cancer, chronic pain, reduced quality of life and early death."

You might say, "That is ridiculous. Donuts are yummy *and* socially acceptable and can't even come close to being as dangerous as smoking!" Yet, according to research presented in the *British Journal of Public Health*, people who carry around too much extra body fat have many more chronic health problems than their smoking counterparts![201] According to the latest studies, being fat reduces the quality of your life and increases your risk of illness and death.[202,203,204,205]

Defining the Problem

In case you are wondering exactly what constitutes being obese these days, researchers who study the effects of body fat on health refer to a person's body mass index (BMI). BMI is an equation that takes into account your weight and height, and through a series of calculations, arrives at a number that should be between 19 and 24.9. According to the National Institutes of Health, if your BMI score is between 25-29.9, you are considered overweight; a score of 30-39.9 denotes obesity while a tally of 40 and over indicates extreme obesity.[206]

The New Obesity Theory

Research on obesity has shifted somewhat over the last few years with the discovery that fat cells have the ability to create loads of inflammatory substances called cytokines. The more fat you gain the more inflammation you create. Some dietary fats (mostly from the omega-6 family) also have the ability to convert into inflammatory messengers, while others (such as GLA and fish oils) have the ability to actually prevent inflammation. Perhaps this is the connection between fats that stop our bodies from burning fat and fats that increase it. Fats that encourage the production of pro-inflammatory messengers (prostaglandins) are found in processed, fried and fast foods, vegetable shortening, corn oil, safflower oil, sesame oil, sunflower oil, margarine and animal fats.

The potential of these fats stimulating inflammatory messengers is quite high – especially if there are insufficient good fats in the diet – and could not only hamper your fat burning goals, but also create an unhealthy cellular metabolism. Fats that encourage the production of anti-inflammatory messengers include omega-3 fats such as cold-water fish, flaxseed and walnuts, as well as gamma linolenic acid or GLA (from borage or primrose oil) and conjugated linoleic acid or CLA. These fats not only enhance fat burning, they also block the production of *Substance P*, a chemical produced in the brain that promotes pain and inflammation.

Stepping in the Right Direction

Let's face it; when you are following an ideal eating and exercise plan, you just feel better. This is not far-fetched. Your body doesn't enjoy performing at sub-par levels; it is meant to perform at optimal levels. But it needs the proper fuel and movement necessary to make this happen.

The challenge with any change is our quest for immediate results. Although it took us years to pack on the excess weight, we want it gone yesterday. As a result, we jump headfirst into fad diets, or blow December's gift money on a gym membership for January. After days or weeks of deprivation or exerting ourselves to the point of exhaustion, we question whether the effort is worth the agony. Sadly, the answer is, more often than not, negative. Before long, we slip back into our old habits.

Instead of looking for instant results, work toward long-term lifestyle changes. You can rev up your metabolism by slowing adjusting your exercise and meal plan. Revisit the earlier chapters on choosing metabolism-boosting foods. For a refresher, remember these guidelines for eating to boost your metabolism:

- Eat five to six metabolically balanced nutrient-dense meals per day – ensuring that you have balanced nutrition entering your body every 2.5 to 3.5 hours to keep blood sugar stable. Nutrient-dense foods are those that are high in vitamins, minerals, essential fats, protein and fibre.
- You should never leave more than 3 hours between meals.
- Make sure that each of your meals is comprised of approximately 40% complex carbohydrates, 30% lean proteins and 30% healthy fats.
- Three of your daily meals should be in solid form and two should consist of liquid nutrition (i.e. you will make your own protein shakes). This adds a highly convenient, yet powerful way to maintain anabolic metabolism and therefore enhance fat loss.
- These meals should consist of between 200 and 500 calories each, depending on your lean body mass and your activity level. The last thing you want to do is feed the fat!
- Ensure that your last meal of the day is at least two hours prior to bedtime and try your best to make that meal no later than 8PM.

Building Metabolic Power

In earlier chapters, I covered the importance of proper exercise to build metabolic-boosting lean muscle: the more lean muscle you have, the better your metabolism at burning body fat every minute of every day. (See Chapter Nine: Exercising your Metabolism.) The most important thing to remember is that your body is designed to *move.* Moving, especially with weight bearing exercise, helps to build muscle. The more muscle, the better your metabolism. The better your metabolism, the more fat you lose. Get moving!

Obesity Prevention Chart

Nutrient	Suggested dosage
Ultimate Protein Energy Shake™	1 serving twice daily
FibreLean™	1 serving, 3 times/day with a full glass of water
Ultimate Longevity (with carnosine)™	1 capsule/day
Ultimate Lean Energy™	3-6 capsules once or twice daily
Multi-vitamin/mineral complex	Follow label instructions
Chromium (as LeafBrand)	400-1,000 mcg/day
X2O mineral sachets	1-2 sachets twice daily in water

twenty

PROSTATE HEALTH

Most men don't even realize they have a prostate, let alone what it does – that is, until it starts acting up. This walnut-sized gland consists of two lobes, enclosed in a layer of tissue and located just below the bladder in front of the rectum. Encircling the urethra (the tube through which urine leaves the body), the prostate is part of the male reproductive system. One of the main roles of the prostate is to squeeze fluid into the urethra as sperm travels through it during ejaculation. This liquid helps to energize sperm and de-acidify the vaginal canal (and you thought you only had to go through sex education once in your lifetime!)

Unfortunately, as men age, the prostate continues to enlarge. Eventually, in a condition called benign prostate hyperplasia (BPH), the tissue surrounding the lobes constricts its outward growth and the gland starts to apply pressure on the urethra, disturbing the flow of urine. As a result, the bladder becomes irritated and bladder walls become inflamed. This causes the bladder to contract when it contains only a small amount of urine, leading to more trips to the bathroom. If urine remains in the bladder, urinary tract infections can occur. Long-term consequences include bladder stones or bladder damage, kidney disease, and incontinence. Although BPH and prostate cancer have not been clearly linked, the diseases occur in a similar population of men. BPH affects more than 50 percent of men past the age of 50. In 2005, experts predict that more than 200,000 new cases of prostate cancer will be detected in the United States, leading to 30,000 deaths.[207,208]

The hormone link

Although men produce both the male hormone testosterone and the female hormone estrogen throughout their lives, levels of active (or free) testosterone decrease as men age. This leaves a higher percentage of estrogen in the blood. Some theories suggest that the imbalance of estrogen increases activity of substances that promote cell growth. Other research indicates that even though testosterone levels fall with age, older men continue to produce dihydrotestosterone (DHT), a substance metabolized from testosterone in the prostate, which may be a factor in its growth. It seems that men who don't produce DHT do not experience prostate enlargement.[I]

Certain genetic factors may also influence the hormonal equation when it comes to prostate cancer. Variations in the DNA found in genes that code for enzymes and hormones involved in the production and metabolism of androgens appear to predict a susceptibility to prostate cancer. Family-based association tests revealed a significant association between prostate cancer and a variation in one of the androgen-metabolizing components.[209] Other 2005 research from the University of California suggests that elevated insulin levels might also lead to prostate cancer. Elevated insulin levels are also a factor in metabolic syndrome.[210] (See Part Two, Condition 17, Metabolic Syndrome.)

Nutrient Factors

Let's start with food groups that are known to aggravate the prostate. Research presented in the *International Journal of Cancer* in 2005 suggests that bread and meat may play a larger roll than once believed. In the study, participants completed food-frequency questionnaires prior to diagnosis of any prostate condition. In combination with diagnoses, results of the surveys suggest that a menu including primarily refined grain products, processed meats, red meat and organ meats contributes to increased prostate cancer risk.[211] Other research supports this conclusion, suggesting that some monounsaturated fatty acids and starches are directly associated with prostate cancer risk, while polyunsaturated fatty acids like linolenic are prostate-protective.[212] (See Chapter 6, Dietary Fats) New research published in the journal *Carcinogenesis* shows that omega-6 fatty acids – found in commercial cooking oils and bakery items – enhance inflammatory messengers within the prostate, fueling prostate growth and cancer risk. In fact, one of the researchers in the study indicated that a diet high in omega-6 fatty acids may actually turn on this cancer cascade, which has been shown to be a common denominator in the growth of prostate, colorectal and some breast cancers.[213] Knowing what you already know about nourishing the metabolism of your cells, this information should make sense to you.

Substantial research also shows the benefits of consuming bright red fruits and vegetables when it comes to protecting prostate health. Watermelons, papayas, pink grapefruits, guava and especially tomatoes are excellent sources of the antioxidant lycopene, the primary carotenoid responsible for the red pigment found in these foods. In a study performed at the Department of Human and Nutrition, University of Illinois, participants consumed tomato sauce (the highest source of lycopene) daily for three weeks before the scheduled removal of their prostate gland. Eating the nutrients in a whole food form was so well accepted by the participants that their blood lycopene levels doubled and their prostate lycopene concentration tripled during this short period. Researchers concluded that the antioxidant power of lycopene might be responsible for the reduction of DNA damage in prostate tissue. The study also found that lycopene might have caused an increase in programmed cell death (apoptosis) of prostate cancer cells. For added benefits, enjoy a drink of green tea with your tomato-based meal. Green tea has also proven to be protective against prostate cancer.[214]

Remembering the hormonal component of prostate health, you will also be interested in the role of phytoestrogens. Phytoestrogen is a fancy term for estrogen-like substances that come from plants. In certain cases, these plant estrogens might be beneficial both for the prevention and for therapy of prostate enlargement. Found in vegetables and fermented soy foods, these plant estrogens help to reduce the impact of male hormones.[215] Just don't overdo the soy foods: too much estrogen is linked to a loss of libido, which is not exactly the greatest trade off. Herbal remedies such as saw palmetto have also been used for centuries for the treatment of enlarged prostate. Studies show that saw palmetto blocks the formation of dangerous testosterones associated with enlarged prostate.[216]

Finally, it is next to impossible to maintain proper cellular metabolism without the metabolic boosting properties found within natural sunlight. One of these properties – vitamin D – has been shown in numerous studies to inhibit the proliferation, invasiveness and metastasis of human prostatic cancer cells. Recently, researchers from the Northern California Cancer Center showed that high sun exposure reduced prostate cancer risk by as much as 65 per cent in a group of men aged 40 to 79 years.[217] Many of us may be facing an inadequate vitamin D status due to the fact that we are constantly warned against sun exposure. Aside from exposing your skin to the sun regularly (this in no way condones tanning!), vitamin D supplementation may be a wise choice.

And exercise?

This chapter wouldn't belong in a book about the value of boosting the metabolism of every cell in your body if I didn't include the benefits of exercise! Researchers from the Harvard School of Public Health analyzed data from the Health Professionals Follow-Up Study for the years 1986-2000. It shouldn't come as any surprise to you that they determined regular vigorous activity could slow the progression of prostate cancer.[218] And to give you another good reason to take the dog for a walk, a study conducted in 2005 found that moderate physical activity reduced risk of developing prostate cancer.[219] When it comes to the healthy metabolism of prostate cells, exercise and proper diet are no longer an option.

Prostate Protection Nutrient Chart

Nutrient	Suggested dosage
Ultimate Prostate™	3 capsules once or twice daily
Ultimate Male Energy™	2 capsules daily
X2O mineral sachets	1-2 sachets twice daily in water

twenty one

SKIN HEALTH: Acne, Eczema and Psoriasis

Unlike beauty, metabolism isn't just skin deep. But your skin sure is a good place to start when it comes to judging the health of your metabolism. Skin is the body's largest organ, and it makes up between 12% and 15% of our total body weight. While we are familiar with the role of skin in touch and absorption, less known are its duties in elimination and detoxification. Skin also has specialized cells to help manufacture vitamin D, regulate body temperature and determine our colour. With so many responsibilities, skin needs our help to stay in top working order. And when things aren't going well inside our bodies, the condition of our skin might be the first clue.

Skin Biology: The really short version

Our skin consists of three major layers: the epidermis (the top layer), the dermis (the second layer) and the subcutaneous layer.

The epidermis is the only visible layer of skin, and it has no blood supply. The epidermis consists of five zones, starting at the bottom with the basal layer. This is where new skin cells are produced, with millions of cells being formed daily. These new cells push upward until they reach the surface. The life cycle of a healthy skin cell is about 28 days. Cells in the basal layer are the only cells that receive nutrients from the dermis, so the skin cells you see are already dead. The dermis is home to collagen (the body's most abundant protein), which provides the structural architecture for skin , and elastin (to give skin "snap"). Sweat glands, necessary for the elimination of toxins as well as a mechanism for regulating body temperature, are also located in the dermis along with oil glands. Blood vessels in the dermis also help to normalize body temperature. The nerve supply in the dermis is responsible for transmitting data like pressure to the central nervous system for interpretation. Underlying the dermis is the subcutaneous layer, which contains fat stores that keep our skin looking "plump" and us looking healthy.[220] Lots of layers, lots of cells and lots of opportunities to benefit from a healthy metabolism.

Feed Your Skin

You learned the importance of having a strong digestive tract in the Digestive Disorders chapter. And you know your body can't use nutrients it can't access. But once your digestion is humming along on all burners, you need to give the cells of your body the nutrients they need not just to survive, but to thrive. Trust me, you'll see the results in the mirror – and the last time I checked, mirrors don't lie! For optimal skin metabolism, be sure that your diet includes an ample supply of the following nutrients:

Vitamin A promotes new cell growth and is necessary for the health of the mucous membranes that line your throat, stomach and mouth.

B Vitamins are necessary for the healthy metabolism of skin and hair. B vitamins also help maintain healthy blood, metabolize carbohydrates and support immune function.

Vitamin C protects against bruising and is essential in the formation of collagen (the support system for your skin.) This vitamin also helps to prevent abnormal blood clotting.

Vitamin D promotes the healthy metabolism of immature skin cells as they travel from the lower layers to the top where you can see them. This vitamin also helps to slow cell division, which is important in conditions like psoriasis and in the prevention of cancer.

Vitamin E is an antioxidant that protects the fats that make up cell membranes. It is also crucial for scar reduction, tissue repair and wound healing.

Selenium is a very powerful free radical scavenger (antioxidant), vital for the integrity of the skin.

Essential Fatty Acids (EFAs) help to create moist, supple skin, while at the same time promoting anti-inflammatory chemicals in the body. Alpha linolenic acid (ALA) from flaxseed, gamma linolenic acid (GLA) from borage oil, and eicosapentaenoic acid (EPA) and docosahexaenoic acid (DHA) from fish oils are the most beneficial. These fatty acids are also essential to the skin's metabolism.

Skin Yardstick

Healthy skin is supple, moist and free of blemishes. Bruises and cuts should heal quickly and minor tears should heal without leaving a scar. Dry, tough, leathery or bumpy skin indicates sun abuse or neglect. While many of us get expression lines as the birthday candles accumulate, (strive for laugh lines instead of frown lines!), there should be minimal skin wrinkling until we have several decades under the belt. Protect the health of your skin by avoiding the harsh rays of the sun and use gentle skin products that don't irritate or dehydrate. Learn about the many toxins in beauty and body care products and be sure to avoid them when you buy. Remember, whatever you place on your skin will eventually end up somewhere in your body – yikes! (See Chapter 3, The Hormone Connection)

Despite what you might be told in advertisements, conditions like acne, eczema and psoriasis cannot be "cured" with a magic cream or medicated ointment. The root of skin conditions often lie within a dysfunctional metabolism at the cellular level and should be treated at the source for thorough and lasting results. Remember that skin is made up of cells, and cells need optimum metabolism to thrive.

Acne

At puberty, our oil (sebaceous) glands become more active. During adolescence, many young men and women might notice an increase in oil production. Oil travels along the hair follicle to leave our skin at the opening known as a pore. Pores can easily become clogged with dirt, skin cells and excess oil causing the follicle to become inflamed. Too much pressure can cause the follicle to rupture, causing surface bacteria, oil and dead skin cells to cascade into the pore. Infection results, and a pimple is born.

While there has been much debate about whether food plays a role in acne, the question has finally been answered by Dr. Loren Cordain from the University of Colorado. Dr. Cordain investigated societies in New Guinea and the Amazon where fruits and vegetables are the major source of food, and discovered that acne is nearly non-existent in these cultures. According to the research, when refined carbohydrates make up the majority of the diet, insulin levels remain permanently high. (Refined carbohydrates are processed, sugar-containing foods, which I like to call "non-foods.") Because our hormones work together, high insulin levels boost levels of other hormones, including the ones that produce oil.[221] You should also eat only organic meat and dairy products if you have acne. Scientists have found a common link between the consumption of dairy foods and the onset of acne. Researchers suggest the acne might be the result of bioactive molecules and hormones found in the milk.[222]

Acne is also an indication that digestive juices in the stomach might be operating at sub-par levels. Give your stomach a break and focus your diet on fruits, vegetables and lean sources of protein. Take smaller bites, and chew your food well. Smaller, more frequent meals might also be helpful. My *Fat Wars* programs have always consisted of three small solid meals and two liquid protein shakes. Other research has also found that acne is a common reaction to stress.[223] Find ways to reduce your daily anxiety levels. And, of course, don't forget the obvious: some products we use on our bodies can clog pores, too. Isolate and eliminate the offending pore-blockers.

Eczema

Eczema, also known as "atopic dermatitis", involves dry, itchy, red and cracked skin and is caused by an immune system response to an allergen. Because the allergy could be to a topical agent or a food, isolating the trigger is often difficult. Talk to your health care provider to determine the best method of identifying your eczema trigger. Eczema is also associated with candida. (See Part Two, Condition 10, Gut Health.) As a result, it's important to supplement with probiotics if you suffer with eczema.[224]

"While the lesions of eczema are the result of water loss, spending too much time in water can exacerbate the problem. Limit your time in a tepid bath to no longer than 15 minutes."

For over 60 years, scientists have known that inflammatory skin conditions like eczema are often the result of a deficiency in the body's ability to properly metabolize essential fatty acids (EFAs). Since then, researchers have found that eczema-sufferers are unable to metabolize the omega-6 fat linoleic acid into the anti-inflammatory gamma linolenic acid (GLA).[225] As a result, if you have eczema, you should supplement with a high quality essential fatty acid formula that contains molecularly-distilled fish oils (for their anti-inflammatory effects) and GLA directly. GLA from organic borage oil is the richest source.[226]

While the lesions of eczema are the result of water loss, spending too much time in water can exacerbate the problem. Limit your time in a tepid bath to no longer than 15 minutes. Soaking in oatmeal can help to relieve the itch, while chamomile ointment is very healing on the lesions. Applying borage oil directly to the skin is also beneficial.

Psoriasis

Often hereditary, psoriasis appears as thick, reddish-brown patches of skin covered with silvery scales. Patches often appear on the back, buttocks, elbows, knees and scalp. In fact, psoriasis is often mistaken for dandruff. Fingernails may also be affected.

Psoriasis is actually a condition of anabolic metabolism gone awry. It involves the rapid uncontrolled reproduction of skin cells. You will recall the healthy skin cycle lasts about a month. For those with psoriasis, the chain of events is much faster, lasting about a week. Ensuring that the diet contains adequate vitamin D is therefore essential for those with psoriasis, as vitamin D slows cell replication.

Although the exact cause is not known, psoriasis has been linked with poor metabolism of fat, so avoid foods that are high in saturated fat, like red meats and fried foods. A deficiency in EFAs also appears to be a trigger. If you have psoriasis, studies show that fish oil supplementation is useful.[227] Mechanisms of the immune system might also play a role in psoriasis, so strengthen your immunity with an antioxidant-rich diet. As with eczema, supplementing with a high quality essential fatty acid formula that contains molecularly-distilled fish oils and GLA is your best option. You can also apply borage oil directly to the lesions to offer relief of symptoms.

Reflections on Metabolism

As you can see from these few examples, a vibrant metabolism affects not only how we feel, but also how we look. Remember, we can't keep the strength of our metabolism a secret just because we are encased in skin!

Skin Protection Nutrient Chart

Nutrient	Suggested dosage
Ultimate Protein Energy Shake™	1 serving twice daily
FibreLean™	1-3 servings/day in a full glass of water
Ultimate Longevity (with carnosine)™	1 capsule/day
Multi-vitamin/mineral complex	Follow label instructions
Vitamin C (mixed ascorbates) with bioflavonoids	500-2,000 mg/day
Vitamin E (as mixed tocopherols)	800 IU daily
Vitamin D3 (Cholecalciferol)	800 IU daily
X2O mineral sachets	1-2 sachets twice daily in water

twenty two

SLEEP AND RESTORE

According to sleep researcher Dr. James Maas from Cornell University, at least half of the American population is chronically sleep-deprived, while the other half has trouble sleeping on most nights. At least 56 percent of the US population complains about daytime drowsiness.[228] It would stand to reason that Canadians are experiencing very similar results.

Aside from the facts that each year a minimum of 100,000 accidents and over 1,500 deaths occur due to people falling asleep at the wheel[229] and that two of the worst nuclear accidents (Chernobyl and Three Mile Island) occurred in the early morning hours while sleep-deprived workers stood watch,[230] researchers are now able to demonstrate that insufficient sleep leads to premature aging, disease and obesity. In fact, fat cell expansion may have more to do with how much time we spend sleeping then we ever imagined!

Numerous studies appearing in peer-reviewed journals over the last few years have shown a direct link between lack of sufficient sleep and ineffective or dysfunctional metabolism. It turns out that even mild sleep deprivation has the ability to cause negative hormonal changes that may in turn alter the complex metabolic pathways responsible for controlling appetite, food intake and the amount of calories we burn each day. A population-based study appearing in the January 2005 edition of the *Archives of Internal Medicine* shows such a link. Researchers had 1,001 patients from four primary care medical practices fill out questionnaires concerning demographics, medical problems, sleep habits and sleep disorders. The conclusion was that insufficient sleep is associated with decreased metabolism, leading to overweight and obesity.[231]

It's all about hormones

Researchers from Stanford University School of Medicine have discovered that the alteration of two key hormones called *ghrelin* and *leptin* may be a key reason as to why tired bodies experience dysfunctional metabolisms and accumulated body fat. Ghrelin is one of the primary metabolic hormones responsible for enhanced hunger signals, while leptin works to suppress appetite. The ground-breaking study using the "real life" conditions of over 1,000 volunteers appears in the December 2004 issue of *Public Library of Science*. It shows that subjects with an average of five hours of sleep per night experienced a 15 percent increase in ghrelin and a 15 percent decrease in leptin than those who slept approximately eight hours on average.[232] These hormonal changes are enough to lead to enhanced hunger and fat gain. Other research, from the University of Chicago, indicates that healthy, sleep-deprived subjects reported a 24 percent increase in appetite, especially for sweet, starchy and salty foods.[233]

More is good

Although studies show a direct correlation between regular loss of sleep and metabolic problems, what about those who lose the occasional night of sleep? Researchers from Penn State University have shed some light on this issue as well. The researchers set out to see how one night of insufficient sleep would affect levels of a potentially harmful immune system protein or cytokine called *Interlukin-6 (IL-6)*. Elevated levels of IL-6 are often seen in the obese, immune-compromised and those suffering from sleep disturbances and excessive daytime sleepiness. Using eight healthy young men between the ages of 20 and 29 over a seven-day period (with only one night of sleep deprivation), the researchers discovered that only *one night* without sleep greatly enhanced blood levels of IL-6 the entire next day. The researchers concluded that a good night's sleep leads to lowered levels of IL-6.[234]

Loss of sleep can also make many of us feel as if we're losing our minds! According to a study appearing in the March 15, 2003 issue of the journal *Sleep*, accumulated sleep-dept does, in fact, rob us of our mental capacity. Researchers discovered that habitually getting six hours of sleep or less per night causes a noticeable decrease in brain performance. The amazing thing was that the chronically sleep-deprived subjects in this study reported feeling "only slightly sleepy," even when psychological testing showed that their performance was at its lowest level.[235]

The news gets worse. Researchers from Columbia University discovered that people who slept less than seven hours each night had a much greater propensity towards obesity. In the study – using subjects aged 32 to 59 – it was found that, compared to those who get a good night's sleep, those who sleep four hours or less each night have a 73 percent greater chance of being obese, those who sleep five hours have a 50 percent greater chance, and those who sleep six hours are 23 percent more likely to be obese.[236]

Research presented in the prestigious journal *The Lancet* shows that sleep deficit can also exert a premature aging reaction on the body by increasing the severity of age-related chronic disorders. The study shows that sleep-deprived people experience a negative effect on hormone function, especially concerning carbohydrate metabolism, and that the effects of sleep shortages are similar to those seen in normal aging.[237]

Early to bed for the little ones

Sleep loss is not experienced solely by adults. In fact, researchers from the Stanford University Sleep Epidemiology Research Center indicate that even school-aged children are generally not obtaining enough restorative sleep, which may be placing them at an increased risk for obesity and diabetes later in life.[238] Researchers from the University of Massachusetts in Boston have also shown a direct correlation among sleep shortage, depressive symptoms and self-esteem issues during adolescence.[239] The problem may be worse than we realize: researchers from Brown University School of Medicine have found that at least one-third of youngsters in kindergarten through the fourth grade may suffer from at least one sleep-related problem.[240]

Sleep Naturally

Do you remember when you were little and you couldn't sleep? If you were like most North Americans, your mom probably heated up some milk and blended in a banana or some honey to help you sleep. The combined effect of an amino acid found in milk (tryptophan), along with the insulin surge created by the blended banana and/or honey probably made you feel sleepy. Tryptophan is an essential amino acid required for the production of one of our most important neurochemicals, serotonin (which promotes balanced moods and feelings of well-being and confidence, reduces stress and anxiety and aids in sleep). People who suffer from inadequate or unbalanced levels of serotonin (through a deficiency of tryptophan) can often experience a broad array of emotional and behavioral problems, including depression (most antidepressants work to elevate serotonin levels), excessive cravings (primarily in the evening), compulsive eating, poor stress adaptation, anxiety, PMS, alcoholism, violent outbursts, sleep problems and obesity. [241,242,243,244]

It is very difficult to obtain high levels of tryptophan in the diet (it is the least common denominator in most protein-rich foods). Tryptophan is in constant competition with other amino acids to enter the brain (and manufacture serotonin), and tryptophan is easily degraded in times of stress. It stands to reason, then, that numerous people across North America may be suffering from serotonin deficiencies. This is one of the primary reasons we may have trouble getting to sleep and staying asleep long enough to repair our metabolisms.

Breakthrough research presented in the prestigious *American Journal of Clinical Nutrition* in May 2005 shows that by consuming an evening milkshake containing the exact alpha lactalbumin found in the whey isolate product, the *Ultimate Protein Energy Shake*™, healthy people could get a better night's sleep and awaken feeling "alive," refreshed and energetic the next morning. Dutch researchers gave 28 healthy young adults with mild sleep problems a milkshake containing the *Ultimate Energy Shakes'* alpha lactalbumin protein, which delivers unsurpassed levels of bioavailable tryptophan to the body and brain. In doing so, the researchers found that the protein caused a 130 percent increase in tryptophan levels before sleep, which seemed to be responsible for a deeper and more restorative sleep. The researchers were also amazed that the participants who had mild sleep problems showed a marked improvement in morning alertness and experienced significantly reduced sleepiness the following day.[245]

The take-home message is simple: if you are interested in losing body fat, gaining muscle, slowing biological aging and maintaining cognitive abilities and an optimal health profile, getting a good night's sleep is no longer optional!

Sleep and Restore Nutrient Chart

Nutrient	Suggested dosage
Ultimate Protein Energy Shake™	1 serving prior to sleeping
Ultimate Anti-Stress™	2-4 capsules prior to sleeping
Melatonin	500 mcg-3 mg prior to sleeping
Magnesium	500 mg prior to sleeping
Ziziphus spinosa seed extract	25 mg prior to sleep

twenty three

THYROID HEALTH: Hypothyroidism

We know that every single cell in our bodies contains its own metabolism that is at work every second of every day, and the integrity of that metabolism affects every other cell in our body. Our body's metabolism is always at work, keeping us alive and thriving – and our thyroid gland is one of the master controllers of that metabolism. The thyroid gland is responsible for controlling the overall metabolism of the body with the aid of its specialized hormones – thyroxin (T4) and triiodothyronine (T3). These hormones are responsible for several metabolic functions, including body temperature, the release of energy from cells and protein synthesis (anabolic metabolism). Even though 93% of the hormone secreted by the thyroid is T4, the great majority of T4 must be converted to T3 (the metabolically active thyroid hormone) in order to enhance metabolism. The great majority of T3 (converted from T4) is manufactured in peripheral (meaning away from the thyroid) tissue, with the liver and kidneys being responsible for approximately 80% of its production. The thyroid must create the rest.

But like many things in life, there can be a breakdown in the system. An important job of the liver, for example, is to keep the bloodstream clean. Too many toxins, unchecked hormones or high fat meals can do a number on the liver, leaving it overworked and not as efficient at creating metabolically active T3. Hormone-producing abnormal tissue growths in the thyroid can also cause the thyroid gland to run amok, leading to over-active thyroid or hyperthyroidism. The thyroid gland eventually becomes exhausted or, as a result of medication used to control over-activity, the thyroid gland starts to produce too little of the metabolically-active hormones, especially T3. And as we age, we tend to experience a decline in this conversion of T4 to active T3.[246] In fact, the late thyroid expert, Dr. Broda Barnes, author of *Hypothyroidism: The Unsuspected Illness*, estimated that over one-third of the adult population was suffering from a subclinical thyroid deficiency or hypothyroidism that robs them of their full metabolic potential.[247]

Indications that your thyroid production might be deficient include uncontrolled weight gain, difficulty losing weight, constant fatigue, depression, cold intolerance, dry skin, brittle fingernails and hair, constipation, heart disease, high cholesterol and poor memory. Because C-cells in the thyroid gland secrete the hormone calcitonin (involved in controlling calcium levels in the blood and bone), thyroid health is also a factor in osteoporosis. The reason the majority of the adult population experiences these low-thyroid symptoms (even when blood tests are negative) is because thyroid blood tests are not always that reliable.

How do you know for sure?

Physicians usually perform a blood test to measure thyroid levels, but there could be problems with the results – or at least the way they are currently perceived. One test is based on the assumption that low-thyroid hormones usually stimulate a rise in the pituitary hormone called TSH in an attempt to manufacture more thyroid hormones. When assessing whether a person is suffering from low thyroid (hypothyroidism), physicians use a TSH laboratory reference range of 0.2 IU/ml to 5.5 IU/ml. The higher the TSH levels, the greater the possibility of low thyroid. Most doctors only prescribe thyroid hormone medication when levels are over 5.5. Other tests, for example, measure the quantity of T4 in the blood, but we already know that T3 is the metabolically active form. So, while you might seemingly have healthy levels of T4 circulating throughout your body, you might be an individual who has trouble making the conversion to T3.

And then there is the fact that you are a unique individual, with your own little system quirks. It's possible that while your blood results are within "acceptable" ranges for the lab, those ranges might not be acceptable or the most efficient for your own body. According to research presented in the July 1995 issue of the journal of *Clinical Endocrinology*, thyroid abnormalities may actually be present when TSH values exceed 1.9, not 5.5. This means that numerous individuals may be undiagnosed with possible thyroid problems. After all, how can you possibly say that a TSH reading as low as 0.2 is the same as

5.5 (a 27-fold difference)? To shed light on the magnitude of the problem, in the summer of 2002, an article appeared in *The Lancet* – that would question the efficacy of all previous thyroid-tests. The article indicated that today's thyroid blood tests do not adequately diagnose low thyroid due to faulty "Reference Ranges."[248] In fact, research from the *Oregon Health Sciences University* in Portland, Oregon discovered that elderly people, who were found to have normal thyroid activity, were later found to have serious thyroid deficiencies after undergoing advanced testing.[249,250,251]

Remember, we don't arrive at serious illness overnight. There is usually a long and winding journey from optimum health to illness and disease. So, if your blood tests show you to be on the low side of normal, you'll want to consider what you can do now to prevent a further decline in metabolic activity. Blood tests don't take these variables into account.

BASAL METABOLIC TEST

If you are experiencing some symptoms of hypothyroid and your lab tests have been negative, try this at-home test to assess your thyroid function:

You'll recall from earlier in this chapter that thyroid function helps to regulate body temperature. Normal resting temperature in the morning is between 97.8 and 98.2 degrees Fahrenheit. Have a thermometer beside your bed and take your temperature first thing in the morning before you get up, while you are lying in bed and as relaxed as possible. Record the results. Take your temperature later in the day as well. Also record the results. Normal active body temperature is 98.6 degrees Fahrenheit. Because hormone fluctuations will alter body temperature, women should start recording body temperature on the first day of menstruation. Take your temperature for anywhere from three to five days. Below average body temperature suggests functional low thyroid status, while higher than average temperature suggests a functional hyperthyroid condition.

Iodine Patch Test

Your thyroid runs on iodine. If you don't have enough iodine in your body, your thyroid can't operate at peak levels. To determine your status, you'll have to buy some tincture of iodine. Before going to bed at night, paint a small patch of iodine on your abdomen. (Wear dark pyjamas!) If the iodine has disappeared into your skin by morning, you are likely deficient in this important thyroid nutrient. A great way to introduce natural iodine into your body is through kelp extracts.

"Eat plenty of iodine-rich foods including seafood and sea vegetables (wakame, kelp, nori), beef, eggs, nuts and seeds. "

Thyroid Therapy

Studies indicate that the metabolically active T3 hormone declines when people go on low-calorie diets. In fact, research from *Rockefeller University*, the Laboratory of Human Behavior and Metabolism in New York showed that even a 10% reduction in bodyweight could cause a reduction in T3 levels. Since numerous obesity problems are associated with ineffective conversion of T4 to T3 and calorie-restrictive dieting itself can worsen the problem (making it next to impossible to lose weight), it makes sense to ensure optimal thyroid health through proper nutrition and exercise – not dieting!

Eat plenty of iodine-rich foods including seafood and sea vegetables (wakame, kelp, nori), beef, eggs, nuts and seeds. Guggulipid from the south Asian mukul myrrh tree is believed to help convert T4 into T3 and increase the uptake of iodine, and L-tyrosine is an essential amino acid for thyroid hormones. Look for L-tyrosine in almonds, avocados, bananas, dairy products, lima beans, pumpkin seeds, and sesame seeds. Your body requires adequate vitamin D to produce thyroid hormone, so be sure to take in about 20 minutes of sunshine daily. You can also get vitamin D from fish and in supplement form.

Stress wreaks havoc on your thyroid health. Be sure to find ways to unwind at the end of the day, perhaps with a walk around the block. You should be able to recite the benefits of exercise right along with me now.

Thyroid Protection Nutrient Chart

Nutrient	Suggested dosage
Ultimate Protein Energy Shake™	1 serving twice daily
Ultimate Anti-Stress™	2-4 capsules prior to sleep
Multi-vitamin/mineral complex	Follow label instructions
Vitamin C (mixed ascorbates) with bioflavonoids	500-2,000 mg/day
Vitamin E (as mixed tocopherols)	800 IU daily
Vitamin D3 (Cholecalciferol)	800 IU daily
X2O mineral sachets	1-2 sachets twice daily in water

appendix I awaken your metabolism

Throughout *Awaken Your Metabolism*, especially in the disorders section, I have mentioned various nutrients that offer special benefits for each related section. Many of these are in the form of synergistic combinations found within special formulas. Due to page count, I could not list all of the ingredients within each formula in every section of the book; therefore I have listed a brief description of each recommended formula below.

Having said this, in my many years in the health industry, I have come to realize that an end product (nutrient formula) is only as good as the quality of its starting ingredients as well as the manufacturing practices. It is therefore imperative to have all bases covered when formulating a high quality nutrient formulation. The criteria I have used as a product formulator over the years is as follows:

- all ingredients within the formula need to be of the highest quality
- all ingredients within the formula need to be research based
- all ingredients within the formula need to work in synergy
- is there a real need for a product like this?
- each new production batch must be third party verified for quality assurance

I guarantee that <u>all of the</u> ***Ultimate*** products are based on sound scientific research and are amongst the top products in the industry (if not the best).

Note: Since I am an avid supporter of the health food stores, most of the recommended supplements in this section can be found throughout Canada at local health food retailers.

The complete line of ***Brad King's Ultimate*** products are distributed (exclusively through health food stores) by:

Canada:

PREFERRED NUTRITION INC.
Hotline: 888-284-9920
Email: www.pno.ca

U.S.A.:
On-line only: www.awakenyour body.com
Website for both Canada and USA: www.awakenyourbody.com

ULTIMATE PROTEIN ENERGY SHAKE™

What is it?

The **Ultimate Protein Energy Shake**™ is designed to supply a perfect matrix of undentaured (undamaged) bioavailable peptides and amino acids to the trillions of cells throughout your body in the least amount of time. The **Ultimate Protein Energy Shake** uses only the highest quality whey isolate and takes it to the next level with a unique cross flow micro filtration method that filters out all impurities and leaves in only the most desirable protein fractions discovered in whey. The result is a *sugar free*, *lactose free*, *casein free*, protein you can actually feel! **Ultimate Protein Energy Shake** contains exceptionally high levels of bioavailable: *alphalactalbumin* (at 33%, it is anywhere from three to four times higher than most other whey isolates on the market), *Branch Chain Amino Acids* (BCAAs) (which are two to three times higher than most other whey proteins), *GMP's* (23%, compared to most other isolates that contain less than 15%) and *glutathione* (GSH) builders.

There are absolutely no artificial sweeteners, high glycemic sweeteners or potentially harmful so-called natural sweeteners used to flavour the protein. We only use 100% natural vanilla or cocoa bean extract along with stevia leaf.

What's its role?

Ultimate Protein Energy Shake™ helps:

- build, repair and replace your bodies cells
- build and repair muscle, skin and bones
- transport various molecules around your body
- transmit messages from brain cells to muscle cells
- enhance your immune system
- provide a backup energy system
- regulate many important metabolic processes
- aid in energy production during the day
- aid in deep restorative sleep during the evening

FIBRELEAN™

What is it?

FibreLean is a 100% natural organic, high-antioxidant, organic fruit and vegetable fibre blend – with absolutely no irritants. Every serving includes five full grams of soluble (58%) and insoluble (42%) *organic* fibres coming from: *Guar Gum*, *flaxseed meal*, *celery*, *apple pectin*, *rhubarb*, *blueberry*, *blackberry*, *cranberry* and *Inulin* (from Chicory extract).

What's its role?

FibreLean™ helps:

- lower excess blood sugar levels
- reduce cholesterol
- fight gallstone formation
- promote healthy weight loss
- stimulate digestion
- detoxify and cleanse the intestinal tract
- enhance beneficial bacteria production
- remove cancer-causing agents from the colon wall
- increase regularity of bowel movements
- reduce varicose veins

ULTIMATE LONGEVITY (with carnosine) ™

What is it?

Ultimate Longevity *with Carnosine* is a one-a-day antioxidant blend (containing: *L-Carnosine, Glutathione (GSH – in reduced form), High-ORAC Blend – pomegranate, apple, elderberry, rosemary, white tea, blueberry, black currant* and *raspberry extracts, ginger extract, thyme extract, rosemary extract* and *bioperine®* – to enhance nutrient absorption), designed to help your body slow down and prevent the often debilitating sugar/protein reactions that take place within almost every one of our trillions of cells. This damaging process is referred to as glycation or glycotoxins and the end stage damage of glycated bodily proteins is an irreversible condition called Advanced Glycation Endproducts or AGE for short.

What's its role?

Ultimate Longevity *with Carnosine*™ helps:

- reduce excess inflammation
- cells revert back to their once youthful state
- protect the cell's genetic code – DNA – from excess oxidation
- enhance the overall antioxidant potential of the body
- increase the powerful antioxidant enzyme – glutathione-S-transferase – used by the body to prevent glycation
- stop the conversion of AGE at a critical point
- prevent muscular fatigue – by removing excess lactic acid
- modulate brain function – by sensitizing neurons to various messages and protecting them from over stimulation
- normalizes heart function, by regulating calcium ions

ULTIMATE HER ENERGY™

What is it?

Ultimate HER Energy is a one-of-a-kind synergistic formulation of naturally proven and scientifically validated herbs, flavonoids, spices and vegetable extracts (containing: *Indole-3-Carbinol (I3C), broccoli extract (containing Sulforaphane), D-Glucarate, citrus bioflavonoids, quercetin, milk thistle, turmeric, holy basil* and *bioperine®* – to enhance nutrient absorption) that help curb the debilitating effects of estrogen-dominance, a condition in which healthful estrogens are pushed out of the way by unhealthy estrogens. New research confirms that hormone metabolism changes with age, so by restoring healthy hormone fidelity, you can once again look, feel and perform in a youthful manner.

What's its role?

Ultimate HER Energy™ helps:

- reduce excess inflammation
- decrease excess body fat – especially in the abdominal region
- protect breast and uterine cells
- fight excess free radicals
- neutralize excess toxins
- support production and longevity of glutathione (GSH) – one of the body's most powerful antioxidants
- enhance overall liver function (aiding in excess hormone removal)
- preserve and build lean muscle tissue
- enhance the body's cellular repair mechanisms
- facilitate the conversion of powerful – cancer-promoting – estrogens, to the safer beneficial 2-hydroxy and 2-methoxy estrogens
- the body adapt to excess stress that causes a depletion in valuable testosterone levels

ULTIMATE MALE ENERGY™

What is it?

Ultimate Male Energy is a one-of-a-kind synergistic formulation of naturally proven and scientifically validated herbs, flavonoids, spices and vegetable extracts (containing: *chrysin*, *stinging nettle extract*, *Indole-3-Carbinol (I3C)*, *broccoli extract (containing sulforaphane)*, *citrus bioflavonoids*, *quercetin*, *turmeric*, *holy basil* and *bioperine®* – to enhance nutrient absorption) that help curb the debilitating effects of lower testosterone and higher estrogen levels that occur as you age, by protecting against estrogen-dominance and restoring testosterone to optimum levels.

What's its role?

Ultimate Male Energy™ helps:

- reduce excess inflammation
- enhance libido
- increase the body's natural production of testosterone
- reduce abdominal fat
- preserve and build muscle tissue
- maintain prostate health
- fight excess free radicals
- the body adapt to excess stress that causes a depletion in valuable testosterone levels
- enhance the body's cellular repair mechanisms
- unlock bound testosterone from its protein transporter – SHBG
- block the activity of the pro-estrogen enzyme – aromatase
- facilitate the conversion of a powerful – cancer-promoting – estrogen, to the safer beneficial 2-hydroxy and 2-methoxy estrogens

ULTIMATE LEAN ENERGY™

What is it?

Ultimate Lean Energy is a "mild stimulating" energy/fat loss formula (containing: *guarana extract, high-EGCG green tea extract, yerba maté extract, forslean® (high-dose Coleus forskholii), gymnema sylvestre extract, cayenne, kelp* and *bioperine®* – to enhance nutrient absorption) that is designed to target five vital areas of metabolism. **Ultimate Lean Energy** may just be one of the safest yet most powerful energy/ fat loss formulas ever developed.

What's its role?

Ultimate Lean Energy™ helps:

- boost your metabolic rate naturally
- inhibit the ability of your fat cells to expand in volume
- decrease the total number of fat cells available for expansion
- enhance natural brain chemicals that break down excess fat
- enhance energy production
- control excess blood sugar and insulin
- reduce sugary cravings

ULTIMATE ANTI-STRESS™

What is it?

Ultimate Anti-Stress formula (containing: *ashwagandha extract, valerian extract, citrus bioflavonoids, lyophilized adrenal tissue, quercetin, rhodiola extract* and *bioperine®* – to enhance nutrient absorption) helps increase the body's ability to deal with excess stress by normalizing adrenal function. By using **Ultimate Anti-Stress** on a regular basis – *in the evening time* – you can aid your body's ability to replenish from the day's unrelenting stress toll, and help avoid adrenal exhaustion or adrenal fatigue.

What's its role?

Ultimate Anti-Stress™ helps:

- normalize adrenal function
- improve overall health
- your body deal with excessive stress
- lower excessive cortisol levels, especially in the evening
- induce relaxation without sedation
- activate fat breakdown instead of muscle tissue
- normalize immunity
- increase mental function
- protect the heart
- place the body in a proper sleep environment

LeafBrand™ will soon be available at Health Food Stores
Email: info@awakenyourbody.com
Website: www.awakenyourbody.com

LEAFBRAND™

What is it?

LeafBrand is the trade name for a line of naturally occurring organically bound, plant-based minerals. LeafBrand minerals are extremely safe for the human body as they are derived 100% from plant sources as opposed to being manufactured from inorganic materials.

What's its role?

LeafBrand™ Chromium helps:

- enhance insulin sensitivity
- lower blood glucose levels
- maintain an optimal metabolism
- maintain and build lean muscle tissue
- improve overall health

Studies indicate that LeafBrand chromium is more bio-available to the body than other forms of chromium, including Chromium Picolinate.

LeafBrand™ Selenium helps:

- provide powerful antioxidant support to the body
- protect the body against carcinogens
- support healthy thyroid function
- protect cell membranes and tissues
- maintain tissue elasticity

SImmunoCare™ is available at Health Food Stores.
Website: www.immuno-care.com

IMMUNO-CARE™

What is it?

Immuno-Care is a one-a-day immune-support supplement that consists of very high potency plant sterols that are synergistically formulated with antioxidants and essential fatty acids to positively balance and support your immune system.

What's its role?

Immuno-Care™ helps:

- strengthen your immune system
- improve overall health

X2O™ is distributed by:

In Canada and the U.S.A.:

XOOMA WORLDWIDE

Web address: www.transformwater.com

X2O

What is it?

X2O is a completely natural coral mineral composite which contains over 70 naturally occurring trace minerals, including calcium and magnesium. These essential minerals (electrolytes) become ionic in water, allowing them to be absorbed quickly and easily by your body. X2O sachets actually make water "wetter" by lowering the surface tension of water molecules.

What's its role?

X2O™ helps:

- your body absorb vitamins and minerals from the foods you eat and the supplements you take
- cleanse the kidneys, intestines, and liver
- protect your body from free radical cell damage
- increase muscle and joint mobility
- increase your oxygen levels
- neutralize harmful acids

NOTES:

NOTES:

Sign up for your
FREE monthly e-letter
with valuable health information
from Brad J. King at:

www.awakenyourbody.com

$2.00 OFF

the purchase of one Ultimate Protein (840g)

REDEEMABLE AT HEALTH FOOD STORES

03-133

BRAD J. KING
M.S., M.F.S.

$1.00 OFF the purchase of any BRAD KING product

REDEEMABLE AT HEALTH FOOD STORES

03-134

BRAD J. KING
M.S., M.F.S.

$1.00 OFF the purchase of any BRAD KING product

REDEEMABLE AT HEALTH FOOD STORES

03-134

Manufacturers Coupon

To the retailer: *Preferred Nutrition* will reimburse the full value of this coupon providing you accept if from your customer on the purchase of the product specified. Failure to send in, on request, evidence that sufficient stock was purchased in the previous 90 days to cover the coupons presented will void coupons. Coupons submitted become property of *Preferred Nutrition*. Store name, account number and staff name must be filled in for redemption.
For redemption, mail to: Preferred Nutrition, 153 Perth St., Acton, ON, L7J 1C9

Limit one coupon per purchase
Not valid with any other offer

Offer valid only in Canada

Manufacturers Coupon

To the retailer: *Preferred Nutrition* will reimburse the full value of this coupon providing you accept if from your customer on the purchase of the product specified. Failure to send in, on request, evidence that sufficient stock was purchased in the previous 90 days to cover the coupons presented will void coupons. Coupons submitted become property of *Preferred Nutrition*. Store name, account number and staff name must be filled in for redemption.
For redemption, mail to: Preferred Nutrition, 153 Perth St., Acton, ON, L7J 1C9

Limit one coupon per purchase
Not valid with any other offer

Offer valid only in Canada

Manufacturers Coupon

To the retailer: *Preferred Nutrition* will reimburse the full value of this coupon providing you accept if from your customer on the purchase of the product specified. Failure to send in, on request, evidence that sufficient stock was purchased in the previous 90 days to cover the coupons presented will void coupons. Coupons submitted become property of *Preferred Nutrition*. Store name, account number and staff name must be filled in for redemption.
For redemption, mail to: Preferred Nutrition, 153 Perth St., Acton, ON, L7J 1C9

Limit one coupon per purchase
Not valid with any other offer

Offer valid only in Canada